NATIONAL UNIVERSITY
LIBRARY   SAN DIEGO

D0788539

NATIONAL UNIVERSITY
LIBRARY   SAN DIEGO

Exposure to Contaminants in Drinking Water

Estimating Uptake through the Skin and by Inhalation

Edited by
Stephen S. Olin

Risk Science Institute
International Life Sciences Institute
Washington, D.C.

CRC Press
Boca Raton   London   New York   Washington, D.C.

ILSI PRESS
Washington, D.C.

Library of Congress Cataloging-in-Publication Data

Exposure to contaminants in drinking water : estimating uptake through
    the skin and by inhalation / prepared by ILSI--International Life
    Sciences Institute, Risk Science Institute Working Group ; Stephen S.
    Olin, editor.
        p.   cm.
    Includes bibliographical references and index.
    ISBN 0-8493-2804-7
      1. Drinking water--Contamination.   2. Biological exposure indices
    (Industrial toxicology)   I. Olin, Stephen S.   II. ILSI Risk Science
    Institute.
    RA591.5.E97   1998
    615.9′02--dc21   98-41145
                                                         CIP

© 1999 International Life Sciences Institute

All rights reserved
No part of this publication may be reproduced, stored in a retrieval system, or transmitted, in any form
or by any means, electronic, mechanical, photocopying, recording, or otherwise, without the prior written
permission of the copyright holder. The International Life Sciences Institute (ILSI) does not claim
copyright in U.S. Government information.

Authorization to photocopy items for internal or personal use is granted by ILSI for libraries and other
users registered with the Copyright Clearance Center (CCC) Transactional Reporting Services, provided
that $0.50 per copy per page is paid directly to CCC, 222 Rosewood Drive, Danvers, MA 01923. Tel:
(978) 750-8400. Fax: (978) 750-4744.

"International Life Sciences Institute," "ILSI," and the ILSI logo image of the microscope over the globe
are registered trademarks of the International Life Sciences Institute. The use of trade names and
commercial sources in this document is for purposes of identification only, and does not imply endorse-
ment by ILSI. In addition, the views expressed herein are those of the individual authors and/or their
organizations, and do not necessarily reflect those of ILSI.

ILSI Risk Science Institute/ILSI Press   CRC Press LLC
International Life Sciences Institute   2000 Corporate Blvd. N.W.
1126 Sixteenth Street, N.W.   Boca Raton, Florida 33431
Washington, D.C. 20036-4810

Library of Congress Catalog Number 98-41145
International Standard Book Number 0-8493-2804-7 (CRC)
International Standard Book Number 1-57881-001-9 (ILSI)
Printed in the United States of America 1 2 3 4 5 6 7 8 9 0
Printed on acid-free paper

About ILSI and the ILSI Risk Science Institute

The International Life Sciences Institute (ILSI) is a nonprofit, worldwide foundation established in 1978 to advance the understanding of scientific issues relating to nutrition, food safety, toxicology, risk assessment, and the environment. By bringing together scientists from academia, government, industry, and the public sector, ILSI seeks a balanced approach to solving problems of common concern for the well-being of the general public.

Headquartered in Washington, D.C., ILSI is affiliated with the World Health Organization as a nongovernmental organization and has specialized consultative status with the Food and Agriculture Organization of the United Nations.

ILSI accomplishes its work through its branches and institutes. ILSI branches currently include Argentina, Australasia, Brazil, Europe, India, Japan, Korea, Mexico, North Africa and Gulf Region, North America, South Africa, South Andean, Southeast Asia, Thailand, a focal point in China, and the ILSI Health and Environmental Sciences Institute, which focuses on global environmental issues. The ILSI Research Foundation includes the ILSI Allergy and Immunology Institute, ILSI Human Nutrition Institute, ILSI Risk Science Institute, and the ILSI Nutrition and Health Promotion Program.

The ILSI Risk Science Institute (RSI) was established in 1985 to advance and improve the scientific basis of risk assessment. RSI serves as a catalyst for consensus on complex scientific issues in risk assessment by facilitating discussion and cooperation among scientists from all sectors.

RSI has joined with the U.S. Environmental Protection Agency in several large cooperative agreements to conduct working groups, workshops, and conferences designed to answer questions in critical areas of risk assessment. This book is the product of a working group organized under such a cooperative agreement with the EPA Office of Water.

Preface

Drinking water is never pure H_2O but always contains at least small amounts of dissolved salts and organic substances. This is true whether the drinking water comes from the tap, from a bottle, or directly from a river or stream. These "other" substances in our water we call *contaminants*, although they may come from natural sources and, in fact, may be essential nutrients such as iron or manganese.

Historically, the primary concern with the safety of drinking water has been microbial contamination, and it is still the principal focus of public water purification systems worldwide. When control of microbial contaminants is compromised, public health inevitably suffers.*

Chemical contamination of drinking water, however, is also of concern, and national and regional regulatory authorities and international advisory bodies, such as the World Health Organization, have set standards and guidelines for acceptable levels of many chemicals in drinking water sources. In the United States, the Environmental Protection Agency's Office of Water, under the Safe Drinking Water Act of 1974 and subsequent amendments, has the responsibility to ensure that the water supplied to the public is safe to drink. This involves establishing enforceable standards, known as Maximum Contaminant Levels (MCLs) for some individual chemicals or groups of chemicals and Maximum Contaminant Level Goals (MCLGs) for a larger number of chemicals. The chemical-specific risk assessments that form the basis of the MCLs and MCLGs must include an assessment of the likely exposure of the population to the chemical if it occurs at a given level in the drinking water. Thus the ability to develop reliable quantitative estimates of exposure is critical to the EPA's task of ensuring the safety of drinking water.

Exposure to contaminants in water may occur when the water is ingested, and this has been assumed to be the major (and perhaps sole) route of exposure in the past. In recent years, however, evidence has been accumulating that for some contaminants, uptake through the skin and/or by inhalation also contributes significantly to the total absorbed dose. The evidence comes both from experimental measurements and from modeling of exposures to contaminants such as chloroform, trichloroethylene, benzene, radon, and others.

There are many uses of tap water that present opportunities for skin contact or inhalation of volatile contaminants or aerosols. Showering, bathing, cooking, washing clothes or dishes, and swimming are examples. For each of these activities, there are a number of parameters that need to be considered when estimating exposures, including physicochemical characteristics of the contaminant, temperature and agitation or dispersion of the water, air flows, frequency and duration of the activity,

---

* G. F. Craun et al. (1993) Conference Conclusions. In: G. F. Craun (ed.), *Safety of Water Disinfection: Balancing Chemical and Microbial Risks.* Washington, D.C.: ILSI Press, pp. 657–667.

permeability of the skin and lung tissue to the contaminant, and other factors. To project the distribution of exposures for a population (or for an individual over a period of time), one must know something about time/activity patterns, characteristics of the buildings in which the potentially exposed population lives and works, demographics, regional and source-specific variation in contaminant levels, and other factors. The major sources of uncertainty and variability in the estimates should also be identified.

Fortunately, these kinds of data are now becoming available for many contaminants and exposure scenarios. Some models have been proposed, but few comprehensive exposure estimates have been attempted to date. It is the purpose of this book to present the current state of the science in this field, to identify and review the available information resources, to evaluate various models and approaches, and to demonstrate the feasibility of developing estimates of the distribution of absorbed doses of contaminants in drinking water through contact with the skin and by inhalation.

The book is the product of a 15-member expert working group convened by the ILSI Risk Science Institute under a cooperative agreement with the U.S. Environmental Protection Agency's Office of Water. Working group members (see Table) were drawn from academia, industry, and government and were selected for their expertise in exposure modeling and measurement, water chemistry, time/activity patterns, dermal and respiratory uptake, and the use of probability distributions in characterizing exposures. Both the structure and the content of the book are the product of the group. Individual chapters and sections were authored by various members of the group and were critically reviewed and discussed extensively by the working group during the book's 2½-year preparation period. The final document represents the consensus views of the group as a whole.

The persistence and dedication of the working group, and in particular the contributions of the authors, are gratefully acknowledged. The encouragement and collaboration of Nancy Chiu, a member of the working group and liaison with the EPA Office of Water, and the support of Jeffery Foran, executive director of the ILSI Risk Science Institute, throughout the project were instrumental in bringing the book to completion. Logistical and secretarial support were ably provided by Diane Dalisera, Andrea Gasper, Kerry White, Mignon Wilson, Mickey Kirchner, and Jennifer Allen. Finally, the working group wishes to express its appreciation to the scientific community conducting the research in this field, to all of those who have provided input and encouragement, and to those whose responsibility it is to estimate and evaluate exposures to contaminants in drinking water, who have patiently awaited the completion of this effort.

Stephen S. Olin

The views expressed here are those of the individual authors and working group members and do not necessarily reflect the views of their respective organizations or the organizations that supported this effort.

ILSI/RSI Working Group on Estimation of Dermal and Inhalational Exposures to Contaminants in Drinking Water

Dr. Annette L. Bunge, Colorado School of Mines

Dr. Nancy Chiu, U.S. Environmental Protection Agency

Dr. Cliff Davidson, Carnegie-Mellon University

Dr. Brent Finley, McLaren/Hart Environmental Engineering

Dr. P. J. (Bert) Hakkinen, The Procter & Gamble Company

Mr. Ted Johnson, TRJ Environmental, Inc.

Dr. John C. Little, Virginia Polytechnic Institute & State University

Dr. James N. McDougal, Air Force Research Laboratory, WPAFB

Dr. Robert R. Mercer, Duke University Medical Center*

Dr. Stephen S. Olin, ILSI Risk Science Institute

Dr. Spyros N. Pandis, Carnegie-Mellon University

Dr. David A. Reckhow, University of Massachusetts

Dr. John Robinson, University of Maryland

Dr. Clifford P. Weisel, Robert Wood Johnson Medical School, UMDNJ

Dr. Charles R. Wilkes, Geomet Technologies, Inc.**

* Current affiliation: National Institute for Occupational Safety and Health.
** Current affiliation: Wilkes Technologies.

Table of Contents

Contributors

Annette L. Bunge, Ph.D.
Chemical Engineering and Petroleum
  Refining Department
Colorado School of Mines
Golden, CO 80401-1887

Nancy Chiu, Ph.D.
Office of Water (WH 586)
U.S. Environmental Protection Agency
401 M Street, S.W.
Washington, D.C. 20460

Cliff Davidson, Ph.D.
Department of Civil and
  Environmental Engineering
Carnegie Mellon University
Pittsburgh, PA 15213

P.J. (Bert) Hakkinen, Ph.D.
The Procter & Gamble Company
5299 Spring Grove Avenue
Cincinnati, OH 45217-1087

Ted Johnson
TRJ Environmental, Inc.
713 Shadylawn Road
Chapel Hill, NC 27514

John C. Little, Ph.D.
Department of Civil Engineering
Virginia Polytechnic Institute and
  State University
Blacksburg, VA 24061

James N. McDougal, Ph.D.
Geo-Centers, Inc.
Air Force Research Laboratory
(AFRL/HEST)
Wright-Patterson AFB, OH 45433-7400

Robert R. Mercer, Ph.D.
National Institute for Occupational
  Safety and Health
1095 Willowdale Road, Mail Stop 2015
Morgantown, WV 26505

Stephen S. Olin, Ph.D.
ILSI Risk Science Institute
1126 Sixteenth Street, NW
Washington, DC 20036-4810

Spyros N. Pandis, Ph.D.
Department of Engineering and
  Public Policy
Carnegie Mellon University
Pittsburgh, PA 15213

David A. Reckhow, Ph.D.
Department of Civil & Environmental
  Engineering
University of Massachusetts
16 C Marston Hall
Amherst, MA 01003-5205

Clifford P. Weisel, Ph.D.
Department of Environmental and
  Community Medicine
Robert Word Johnson Medical
  School — UMDNJ
681 Frelinghuysen Road
Piscataway, NJ 08855-1179

Charles R. Wilkes, Ph.D.
Wilkes Technologies
10126 Parkwood Terrace
Bethesda, MD 20814

1 Introduction

Stephen S. Olin

Exposure assessment, defined by EPA as "the determination or estimation (qualitative or quantitative) of the magnitude, frequency, duration, and route of exposure" for a chemical (U.S. Environmental Protection Agency, 1992), is a key component of any risk assessment for an individual or for populations (National Research Council, 1983; National Research Council, 1994). Human exposures to chemical substances occur primarily by ingestion, by inhalation, and by contact with the skin. An individual's exposure to a particular chemical depends on many factors, such as the occurrence and distribution of the chemical in various media (e.g., air, water, soil), the levels at which the chemical occurs, and the activity patterns and lifestyle of the individual.

Drinking water typically contains low levels of many chemical contaminants and is a potential source of exposure to these contaminants. For most chemicals, it has been assumed for practical purposes that the major route of exposure to drinking water contaminants is by ingestion. However, drinking water (or tap water) is used not only for drinking but also for cooking, showering, bathing, washing, laundering, cleaning, and so forth. As a result, for many drinking water contaminants, there is the potential for exposure and uptake not only by ingestion but also through contact with the skin or by inhalation.

Exposure by ingestion is routinely estimated from the volume of water consumed by an individual (approximately 1 to 2 liters per day) and the concentration of the contaminant in the water. If information is available on the fraction of the contaminant absorbed in the gastrointestinal tract, that information can be used in estimating absorbed dose; if no such information is available, 100% uptake is typically assumed. For exposures to water contaminants by skin contact or inhalation, estimation of the absorbed dose is much more complicated and has been attempted for only a few contaminants to date. It is the purpose of this book to review the state of the science in this field and to present methodologies and data resources for characterizing population distributions of absorbed dose of water contaminants through the skin and by inhalation.

Figure 1.1, developed by the ILSI/RSI working group, is a flow diagram depicting the kinds of data to be considered in assessing the exposure of a population to a contaminant in water. The central focus of the diagram is the **CONTACT PATHWAYS** box, in recognition of the fact that individuals in a population may come in contact with a contaminant in their drinking water in a number of ways that depend on the physicochemical properties of the chemical contaminant (**Chemical-specific**), particularly its volatility (**Volatile/nonvolatile**), and that may involve direct use of

1-57881-001-9/99/$0.00+$.50
© 1999 by International Life Sciences Institute

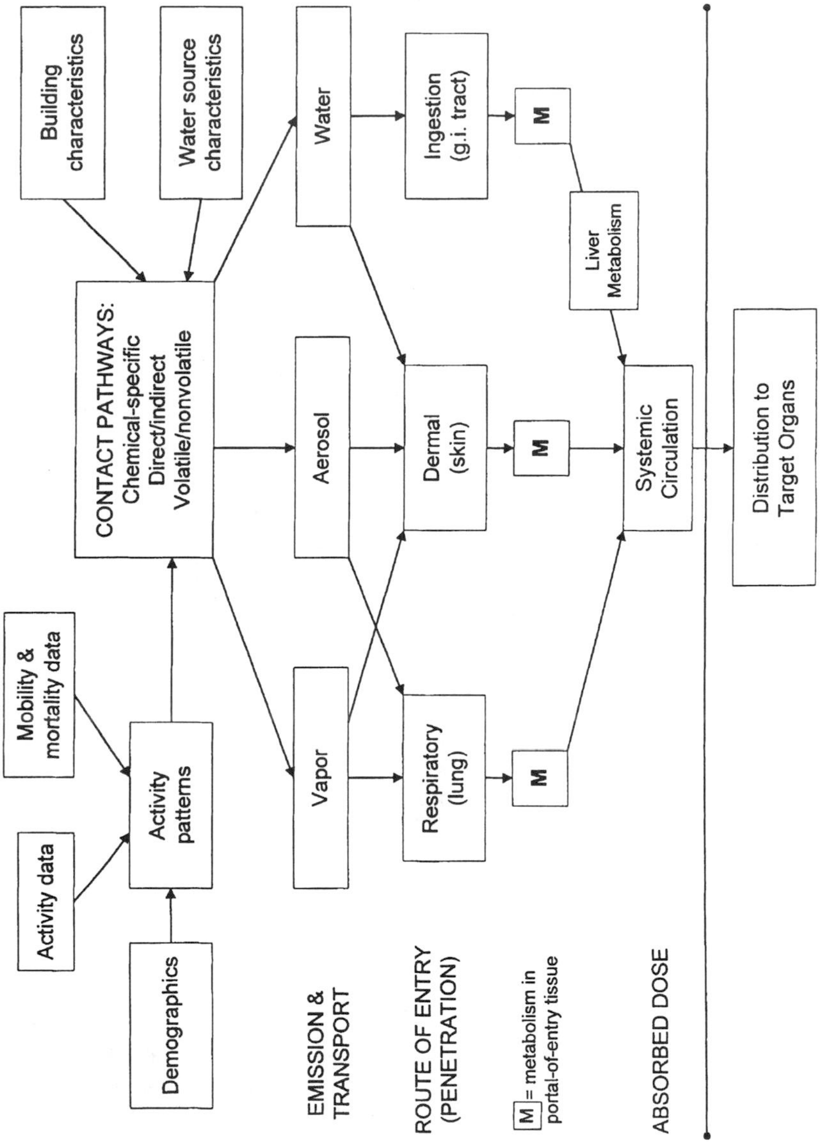

FIGURE 1.1   Exposure assessment of contaminants in drinking water.

water by the individual or indirect exposure (e.g., to vapors or aerosols generated by other occupants of the residence) (**Direct/indirect**). The significance of various contact pathways for a given contaminant is determined by **Water source characteristics** (prevalence and levels of the contaminant, treatment/distribution system effects), **Building characteristics** (volume, ventilation), and **Activity patterns** of the occupants. Activity patterns are how the individuals in the population spend their time, especially their use of water in showering, cooking, washing dishes, laundering, etc. and how much time they spend in the various microenvironments within the residence. (The working group recognized that there is potential for exposure to contaminants in tap water outside the home environment, but chose to focus on residential exposures to illustrate the methodologies.) Some **Activity data** are available on the U.S. population or subpopulations, but **Demographics** and **Mobility & mortality data** (the mobility of the population in moving from one residence to another and mortality as a determinant of residence duration) also should be considered.

The flow diagram continues on by recognizing that exposure to a contaminant may occur by volatilization of the contaminant from water as a gas or vapor (**Vapor**), generation of an aqueous aerosol in activities involving water (**Aerosol**), or by direct contact with water (**Water**). These are seen as categories (or means) of EMISSION & TRANSPORT of the contaminant that may lead to exposure by one or more of the three principal routes: **Respiratory, Dermal**, or **Ingestion**. Exposure to a contaminant as a vapor will be mainly by the respiratory or dermal routes. The same is true for exposure to contaminants in aerosols. Exposure to contaminants by direct contact with water itself comes by ingestion or by the dermal route.

These exposure routes are also the ROUTES OF ENTRY (PENETRATION) into **Systemic Circulation** within the body, characterized by the ABSORBED DOSE of the contaminant. The diagram, however, also recognizes the possibility that metabolism (**M**) occurs in the portal-of-entry tissue (respiratory tract, skin, gastrointestinal tract), thereby reducing the systemic dose of the contaminant and generating metabolites. Finally, the flow diagram acknowledges that absorbed dose is not equivalent to dose at the target organ, and that pharmacokinetics and pharmacodynamics must be considered in integrating exposures by various routes to arrive at a suitable description of dose to the target organ (or organs). Indeed, for some contaminants, even the qualitative nature of the biological effect can depend on the route of exposure. The horizontal line below **Systemic Circulation** indicates that this working group considered exposure only to the point of absorbed dose, especially absorbed dose through the skin or the lungs, leaving perhaps for another working group the issue of an integrated total exposure assessment for all routes.

The document begins (Chapter 2, Contaminant Characteristics) with a discussion of the diversity of chemical contaminants identified in drinking water supplies. While no attempt is made to present a comprehensive list of contaminants, the major chemical classes (based on prevalence and concentrations and on toxicity concerns) are reviewed, with emphasis on the specific contaminants for which U.S. drinking water standards have been proposed or established. Physicochemical properties that significantly influence the occurrence of these contaminants in microenvironments as vapors or in aerosols and the likelihood of exposure and uptake by inhalation or through the skin are also discussed.

Chapter 3, Exposure Characteristics, discusses the role of demographic data in exposure assessment for water contaminants and reviews the available sources of these data for the U.S. population. The domestic activities that involve water are introduced and the data available for characterizing water-related activity patterns for a population are summarized. Data sources for relevant residential building characteristics (type/subtype, volume, ventilation) are reviewed, and examples of data tables are presented. Also, the significance of water source characteristics in estimating contaminant exposures is discussed, including water source quality, treatment, and distribution. Data gaps and research needs are identified in each case.

In Chapter 4, Developing Exposure Estimates, the methods that have been developed and applied in estimating exposures to drinking water contaminants are reviewed and evaluated. Methods are considered for (1) volatiles released from water to indoor air, (2) aerosols and water droplets, and (3) direct contact with water. A procedure for modeling of indoor microenvironments and estimating exposures of individuals in these environments is presented.

Chapters 5, Respiratory Uptake, and Chapter 6, Dermal Uptake, discuss in some detail the estimation of uptake of a contaminant ("absorbed dose" in Figure 1.1) following exposure by inhalation or by skin contact.

Chapter 7 presents three case studies that illustrate the application of the methods and approaches discussed in the first six chapters. Exposures are estimated using an indoor air quality and human exposure model (MAVRIQ: Wilkes et al., 1996) specifically adapted for this purpose. For the calculations, the case studies assume a uniform 1 mg/L concentration of chloroform, methyl parathion, and chromium in the tap water. A single-family home is modeled with two residents and typical room volumes and air flows. Distributions of absorbed doses of each chemical are calculated for uptake through the skin and by inhalation, separately for each route, and combined. These case studies for the three chemicals are presented, not to serve as comprehensive population-based exposure assessments ready for use in risk assessment and regulatory decision-making, but to illustrate the methodologies and provide a framework for discussion of practical issues and applications.

The working group concluded that methods are now available for modeling of microenvironments, such as the MAVRIQ model used in these case studies, that offer flexible and reasonably robust approaches to estimating the extent of exposure and uptake of contaminants in drinking water through the skin and by inhalation. Knowledge gaps and data needs remain, as noted in several of the chapters, and it is the hope of the working group that this book will catalyze and encourage further research in this field.

REFERENCES

National Research Council (1983) *Risk Assessment in the Federal Government: Managing the Process*, National Academy Press, Washington, D.C.

National Research Council (1994) *Science and Judgment in Risk Assessment*, National Academy Press, Washington, D.C.

U.S. Environmental Protection Agency (1992) Guidelines for exposure assessment. *Federal Register* 57(104):22888–22938 (May 29).

Wilkes, C. R., M. J. Small, C. I. Davidson, and J. B. Andelman (1996). Modeling the effects of water usage and co-behavior on inhalation exposures to contaminants volatilized from household water. *J. Exposure Anal. Envir. Epidem.*, 6(4):393–412.

2 Contaminant Characteristics

David A. Reckhow and Stephen S. Olin

CONTENTS

2.1 INTRODUCTION

Literally thousands of chemical compounds have been identified in drinking waters in the U.S. and around the world over the past several decades. In this chapter, we will first consider briefly some of the physical and chemical characteristics of these compounds that will be important in estimating their uptake from water by skin contact and by inhalation. We will then identify the chemicals for which drinking water standards and guidelines have been established and will introduce some of the best-studied classes of chemicals that often appear as contaminants in drinking water supplies.

2.2 PHYSICAL AND CHEMICAL PARAMETERS

Estimating the extent of a population's exposure to a contaminant in drinking water is a complex process. One of the keys is having a good understanding of the inherent properties of the contaminant itself. Volatile contaminants, for example, may be released into the air during the routine use of water in the home and may be inhaled or transported to other locations, whereas exposure to nonvolatile contaminants is more likely to occur through skin contact with water or inhalation of aerosols. Volatility is a function of fundamental characteristics of the contaminant like chemical structure and functionality and molecular weight, and secondary characteristics like partition equilibria and dissociation and complexation phenomena. Respiratory and dermal uptake also are determined, to a considerable extent, by the physical and chemical properties of the contaminant as presented in aqueous solution, aerosol, or vapor.

1-57881-001-9/99/$0.00+$.50
© 1999 by International Life Sciences Institute

Some of the contaminant-specific parameters that will be used in subsequent chapters are introduced in the paragraphs that follow. Standard reference compilations of these values are mentioned where appropriate. For compounds whose physical properties have not been adequately studied, methods for estimating these properties may be available (see, for example, citations in Chapters 4, 5, and 6 and other works such as the *Handbook of Chemical Property Estimation Methods* (Lyman et al., 1990)).

Partition equilibria are of critical importance in assessing a chemical's tendency to migrate from one environmental medium to another (e.g., water to air) and also in determining the permeability of human tissue to the chemical. Many types of partition coefficients have been reported for drinking water contaminants, but their proper use requires not only an understanding and assessment of their nature and purpose and the quality and reliability of the data, but also (mundanely) careful attention to dimensions and units. For well-studied chemicals, multiple values for a given parameter may be available in the published literature, and a selection among values may need to be made based on the intended application, data source, and an evaluation of reliability.

Among the partition equilibrium values of interest are *solubility in water* (the maximum dissolved concentration of a chemical in water at a given temperature) and *vapor pressure* (a measure of the volatility of a chemical, expressed as the pressure exerted by the vapor in equilibrium with its liquid or solid phase at a given temperature). Values for most water contaminants of interest can be found in general sources such as *Lange's Handbook of Chemistry* (1992) and the *Handbook of Environmental Data on Organic Chemicals* (Verschueren, 1996). A recently completed major work, *Illustrated Handbook of Physical-Chemical Properties and Environmental Fate for Organic Chemicals, Vol. I–IV* (Mackay et al., 1992a, 1992b, 1993, and 1995), presents a very complete picture of the physical properties and partition data of these compounds.

The partitioning of a dissolved gas between a dilute aqueous solution and the gas phase is described by Henry's Law, which says that at equilibrium the concentrations in the two phases are proportional. The temperature-dependent proportionality constant, the *Henry's Law constant*, is described in Chapter 4, and values for a number of chemicals are given in Table 4.1; the larger the Henry's Law constant, the greater the proportion of the chemical in the vapor phase at equilibrium. Large compilations of Henry's Law constants are available in Mackay and Shiu (1981) and in Mackay et al. (1992–1995) cited above. If necessary, the Henry's Law constant may be estimated from the ratio of the vapor pressure and the water solubility, although the accuracy is often quite variable because of the lack of reliable data on water solubility.

In practical exposure scenarios, conditions of water use are often dynamic rather than static (e.g., in showering), and in these situations *mass-transfer coefficients* (see Chapter 4) are more useful than equilibrium constants like the Henry's Law constant in estimating vapor-phase exposures to contaminants in water.

The *octanol–water partition coefficient* of a chemical reflects its lipophilic or hydrophilic character and is related to the likelihood that the chemical will be taken up from an aqueous solution by the skin (see Chapter 6). Octanol–water partition coefficients also have been used for many years in modeling environmental fate and transport and other applications, and data have been reported for many compounds (Leo et al., 1971; Hansch et al., 1995; and Verschueren, 1996).

Acid/base dissociation is an important consideration for organic acids and bases in determining volatility and skin permeability. The equilibrium between an acid and its dissociated form ($RCO_2H \rightleftarrows RCO_2^- + H^+$) is expressed by an acid dissociation constant, and the extent of dissociation is, of course, pH-dependent. In its dissociated form, an organic acid in aqueous solution generally is not volatile nor will it readily penetrate the skin. The same is true of an organic base in its protonated form. Acid/base dissociation constants are compiled in many general reference sources, such as the *CRC Handbook of Chemistry and Physics* (1995), *Lange's Handbook of Chemistry* (1992), and the *Handbook of Environmental Data on Organic Chemicals* (1996).

Complex formation can dramatically alter the physical and chemical behavior of metals and organic ligands in solution. The ability of a substance to form stable complexes can change its effective charge, its hydrophilic or lipophilic character, its molecular size, and other properties that affect volatility and uptake through the skin. Many transition metals (see Section 2.3.1) can complex with ligands commonly present in source waters, ranging from ammonia and hydroxide to simple amines and amino acids, carboxylates, sulfides, and phenoxides to the much larger humic and fulvic acids. These processes are discussed in standard texts on water chemistry (e.g., Stumm & Morgan, 1996), and stability constants for some of these complexes have been measured (Smith & Martell, 1974).

Reactivity of a chemical affects its persistence in the environment (and therefore its exposure potential), as well as its ability to be transformed into other substances that also may need to be considered in an exposure assessment. Here, by reactivity we mean the breaking and forming of covalent bonds. Disinfection processes (chlorination, chloramination, ozonation) initiate many complex chemical reactions with organic substances in treated waters, and some of the reaction products are described in Section 2.3.2.

In aqueous media *hydrolysis* is always available as a degradation process for susceptible moieties. Rates of hydrolysis depend on chemical structure, pH, temperature, and the presence of other catalysts (e.g., certain metal ions). Data for many industrial chemicals and common water contaminants were reviewed by Mabey & Mill (1978) and more recently in the Mackay et al. series (1992–1995). Many of the halogenated disinfection byproducts will undergo hydrolysis reactions (see Trehy & Bieber, 1981), and hydrolysis equilibria and rates for the aliphatic aldehydes have been reported by Hoffmann and co-workers.

Oxidation/reduction reactions also may occur as a result of the oxidizing properties of disinfection reagents, from air oxidation, and by interactions with metal ions in solution. Although many drinking water contaminants are relatively stable with respect to redox processes, some are not and these reactions are reviewed in the Mackay et al. series (1992–1995) and in general texts on water chemistry. Reactions of some ozonation byproducts with chlorine residuals have been studied (McKnight & Reckhow, 1992), as have reactions of chlorination byproducts with sulfur nucleophiles (Croué & Reckhow, 1989).

Photolysis of light-sensitive water contaminants generally requires prolonged exposure to sunlight or exposure to high-intensity ultraviolet light (as in ozone/UV water treatment systems). Sunlight can initiate photodegradation reactions of some highly halogenated chemicals (especially those containing bromine), chemicals with aromatic rings or conjugated systems, and compounds containing other photolabile functional groups (e.g., nitro, azo, certain carbonyls). Air oxidation processes also are often catalyzed by the ultraviolet portion of sunlight. Estimation of the rates of environmental photodegradation of contaminants in water, however, is difficult.

2.3 THE CONTAMINANTS

Of the thousands of chemical compounds that have been identified in U.S. drinking waters, most are probably of little consequence to human health, either because they are present at very low concentrations or because they are not very toxic to humans. The U.S. Environmental Protection Agency (EPA) has set Maximum Contaminant Levels (MCLs), Maximum Contaminant Level Goals (MCLGs), and/or Secondary Maximum Contaminant Levels (SMCLs) for about one hundred organic chemicals, metals, and other inorganic species and radionuclides (Table 2.1), representing chemicals that may have a significant impact on water quality or present health concerns. Guidelines for short-term and longer-term levels have been developed by EPA as Drinking Water Health Advisories for another hundred or so substances (U.S. EPA, 1996). States may also establish standards for their drinking water supplies, as long as they are at least as stringent as the federal standards.

The World Health Organization also has issued guideline values for over a hundred chemicals in drinking water. These values are not mandatory limits but may be used by national governments and agencies to set limits for their jurisdictions, taking into account the national or local context of "environmental, social, economic, and cultural conditions" (WHO, 1993).

In this section, several major classes of drinking water contaminants will be introduced and briefly described. Emphasis is given to those classes of chemicals for which standards and guidelines have been developed based on health concerns. In the U.S., the disinfection byproducts and volatile organic compounds, especially solvents, have received the most national attention. Other chemicals, such as certain metals in the water supply or contaminants entering the groundwater from a hazardous waste site, may be of concern at the regional or local level, but it is the disinfection byproducts and volatile organics that have dominated the national scene for many years. More recently, selected individual chemicals including radon, mercury, and arsenic have become the focus of increased efforts to determine safe levels in drinking water.

In the following paragraphs, the inorganics are introduced first for simplicity, and the organics then are presented as (1) disinfection byproducts from the four principal disinfectants in current use (chlorine, chloramines, chlorine dioxide, and ozone), and (2) other classes of organics (chlorinated solvents, pesticides, miscellaneous). The purpose here is simply to describe the class in general terms and identify some of the main chemicals in each class that may occur in water supplies. The

TABLE 2.1
U.S. Drinking Water Standards

Parameter[1]	MCL[2,a]	SMCL or MCLG[3]
	Inorganic Species	
Metals[b]		
Al		0.05–0.2[s]
Ag		0.1[s]
Ba	2	2
Be	0.004 (draft)[d]	0.004 (draft)
Cd	0.005[c]	0.005
Cr (total)	0.1[c]	0.1
Cu	1.3	1[s]
Fe		0.3[s]
Hg (inorganic)	0.002[c]	0.002
Mn		0.05[s]
Na	m&r[5]	
Pb	0.015[e]	0
Sb	0.006[d]	0.006
Tl	0.002[d]	0.0005[d]
Zn		5[s]
Radionuclides		
Beta (mrem)	4	(0)
Gross Alpha (pCi/L)	15	(0)
^{226}Ra + ^{228}Ra (pCi/L)	5	(0)
Rn (pCi/L)	(300)	(0)
U (pCi/L)	(20)	(0)
Non-Metals		
As[4]	0.05 (draft)	
Se	0.05[c]	0.05
Nitrogen Species		
Nitrate + Nitrite (as N)	10[c]	10
Nitrite (as N)	1[c]	1
Cyanide	0.2[d]	0.2
Sulfur Species		
Sulfate	500[g]	250[s]
Halogen Species		
Chloride		250[s]
Fluoride	4[h]	2[h,s]
Chlorine	(4)	(4)
Chloramine (as free chlorine)	(4)	(4)
Chlorine dioxide	(0.8)[i]	(0.3)[i]
Chlorite	(1)[j]	(0.08)[j]
Bromate	(0.01)[j]	(0)[j]
Asbestos (fibers/L, >10 μm length)	7×10^{6} [c]	7×10^{6}

TABLE 2.1 (continued)
U.S. Drinking Water Standards

Parameter[1]	MCL[2,a]	SMCL or MCLG[3]
	Organic Species	
Disinfection Byproducts[6]		
Total Trihalomethanes[7]	0.10	
Total Haloacetic Acids[8]	(0.06)	
Chloral hydrate	(0.06)	(0.04)
Water Treatment Chemicals		
Acrylamide	No MCL[c,f]	0
Epichlorohydrin	No MCL[c,f]	0
Volatile Organic Compounds (VOCs)		
Benzene	0.005	0
Carbon tetrachloride	0.005	0
Chlorobenzene	0.1[c]	0.1
o-Dichlorobenzene	0.6[c]	0.6
p-Dichlorobenzene	0.075	0.075
1,2-Dichloroethane	0.005[c]	0
1,1-Dichloroethylene[9]	0.007	0.007
cis-1,2-Dichloroethylene	0.07[c]	0.07
trans-1,2-Dichloroethylene	0.1[c]	0.1
1,2-Dichloropropane	0.005	0
Ethylbenzene	0.7[c]	0.7
Hexachlorobenzene	0.001[d]	0
Hexachlorocyclopentadiene	0.05[d]	0.05
Methylene chloride[10]	0.005[d]	0
Styrene	0.1[c]	0.1
Tetrachloroethylene[11]	0.005[c]	0
Toluene	1.0[c]	1.0
1,2,4-Trichlorobenzene	0.07[d]	0.07
1,1,1-Trichloroethane	0.2	0.2
1,1,2-Trichloroethane	0.005[d]	0.003
Trichloroethylene (TCE)	0.005	0
Vinyl Chloride	0.002	0
Xylenes	10[c]	10
Pesticides		
Alachlor	0.002[c]	0
Aldicarb	0.007 (draft)[12]	0.007 (draft)[12]
Aldicarb sulfoxide	0.007 (draft)[12]	0.007 (draft)[12]
Aldicarb sulfone	0.007 (draft)[12]	0.007 (draft)[12]
Atrazine	0.003[c]	0.003
Carbofuran	0.04[c]	0.04
Chlordane	0.002[c]	0
2,4-D	0.07[c]	0.07

TABLE 2.1 (continued)
U.S. Drinking Water Standards

Parameter[1]	MCL[2,a]	SMCL or MCLG[3]
Dalapon	0.2[d]	0.2
Dibromochloropropane	0.0002[c]	0
Dinoseb	0.007[d]	0.007
Diquat	0.02[d]	0.02
Endothal	0.1[d]	0.1
Endrin	0.002[d]	0.002
Ethylene dibromide	0.00005[c]	0
Glyphosate	0.7[d]	0.7
Heptachlor	0.0004[c]	0
Heptachlor epoxide	0.0002[c]	0
Lindane	0.0002[c]	0.0002
Methoxychlor	0.04[c]	0.4
Oxamyl (Vydate)	0.2[d]	0.2
Pentachlorophenol	0.001	0
Picloram	0.5	0.5
Simazine	0.004[d]	0.004
Toxaphene	0.003[c]	0
2,4,5-TP	0.05[c]	0.05
Miscellaneous Organics		
Benzo[a]pyrene[13]	0.0002[d]	0
Di(2-ethylhexyl)adipate	0.4[d]	0.4
Di(2-ethylhexyl)phthalate	0.006[d]	0
PCBs	0.0005[c]	0
2,3,7,8-TCDD	3×10^{-8} [d]	0

[1] All concentrations in units of mg/L unless otherwise indicated. Numbers in parentheses indicate proposed standards where no current standard exists.

[2] General notes on MCLs:

  a = The MCLs (maximum contaminant levels) are the maximum permissible levels in water that is delivered to users of a public water system.

  b = An MCL (and MCLG) for Ni, a Phase V contaminant, was proposed but was later remanded.

  c = Phase II contaminants published in January 1990, effective 1992.

  d = Phase V contaminants published July 1992, effective January 1994.

  e = These are action levels (ALs) instead of MCLs. This is part of the "Lead and Copper Rule" that was promulgated on 7 June 1991. Both lead and copper ALs must be met in at least 90% of samples.

  f = Control through a treatment technique requirement, rather than a formal standard.

  g = Proposed sulfate rule; also includes a 500 mg/L MCLG.

  h = Under review (October 1996).

  i = Tentative.

  j = Listed for regulation.

[3] All listings under this column represent nonenforceable standards. The SMCLs (secondary maximum contaminant levels) are recommended levels for aesthetic considerations. They are all marked with a superscript "s". The MCLGs (maximum contaminant level goals) are levels where no adverse health effect would occur with a margin of safety.

TABLE 2.1 (continued)
U.S. Drinking Water Standards

[4] A new arsenic rule is anticipated. NPDWR proposal should be in 2000 with promulgation by 2001. An MCL in the range of 0.002 to 0.020 mg/L is expected.

[5] No MCL; only monitoring and reporting required.

[6] A wide range of disinfection byproducts are under consideration for future MCLs. These include the haloketones, haloacetonitriles, halophenols, chloropicrin, cyanogen halides, and low-molecular-weight aldehydes.

[7] Proposed Disinfectants/Disinfection Byproduct Rule includes an MCL of 0.080 mg/L for total trihalomethanes and MCLGs of 0 mg/L for chloroform, bromoform, and bromodichloromethane, and 0.06 mg/L for chlorodibromomethane.

[8] This is the MCL found in the proposed Disinfectants/Disinfection Byproduct Rule.

[9] Also known as vinylidene chloride.

[10] Also known as dichloromethane.

[11] Also known as perchloroethylene or PCE.

[12] The MCL (or MCLG) for any two of these three chemicals (aldicarb, aldicarb sulfoxide, aldicarb sulfone) should not exceed 0.007 mg/L.

[13] PAHs were intended to be regulated under the Safe Drinking Water Act Amendments of 1986, but action has been taken only on benzo[a]pyrene.

---

reader is referred to current texts on water chemistry and reviews (e.g., Richardson, 1998) for more comprehensive discussions.

2.3.1 INORGANICS

The inorganic chemicals for which drinking water standards have been established include a number of metals, several nonmetallic inorganic species, and some radionuclides (see Table 2.1). Many contaminated groundwaters and some surface waters contain elevated concentrations of one or more of these species. Among the more frequently reported members of this class are lead, chromium, and arsenic.

Metals in drinking water may be grouped into three classes: (1) those that are present in the raw water and that are incompletely removed through treatment, (2) those that are purposely added during treatment, and (3) the corrosion byproducts. Aluminum is the principal metal in the second group. Members of the third group include lead, copper, and to a lesser extent cadmium and chromium. Except for mercury, the transition metals are generally less volatile than other water contaminants, although they may be transported in aerosols or dusts. Metals in drinking water are almost invariably present as ions, and oxidation state and the formation of complexes and organometallic compounds are key considerations in exposure and risk assessments for this class.

Although *aluminum* is one of the most abundant elements in the earth's crust, it is rarely present at high concentrations in water. This is due to its strong tendency to form insoluble hydroxides. For this reason, addition of aluminum in the form of alum for the treatment of water does not generally lead to elevated levels of the metal in finished water. However, problems may occur when the pH of the water drops too low (e.g., 5 or below). Under these conditions, aluminum hydroxide precipitates will not form to as great an extent, and aluminum solubility rises sharply.

Antimony exists in two oxidation states, (+III) and (+V). Antimony is normally present at very low concentrations in water, and as a result little is known about its speciation. Of the 50 elements studied by the U.S. Geological Survey in soil and rock samples, antimony was the third least abundant, with an average composition of 0.5 ppm (Shacklette and Boerngen, 1984). Antimony may become methylated in aquatic environments, which can substantially increase its solubility (Andreae, 1983).

Beryllium is not widely found in the natural environment. The presence of high concentrations of this metal suggest contamination from industrial wastewater, especially nonferrous metal manufacturers. The atmosphere may also serve as an important beryllium source for natural waters (U.S. EPA, 1987).

Although *cadmium* is a regulated drinking water contaminant, water is rarely a major route of exposure. Instead, food and incidental ingestion of dust are thought to be much more important. However, in some instances, significant amounts of cadmium have been found to leach from plumbing fixtures.

Chromium exists in water in either the +III or the +VI oxidation state. Although Cr(+III), the most common form of chromium, is an essential nutrient for the human diet, Cr(+VI) is considered to be carcinogenic, at least by inhalation (IARC, 1990), and high atmospheric levels of Cr(+VI) have generated human health concern. Chromium will adsorb to metal oxides and form strong chelates in natural waters. As a result, it is primarily found in water in a particulate form, with some soluble chromium complexes. The mean U.S. chromium level for river water has been estimated at 10 µg/L (Eckel & Jacob, 1988). Concentrations in tap water have been found to be lower (1.8 µg/L average, Greathouse & Craun, 1978), but corrosion of plumbing fixtures can drive these up. Some contaminated groundwaters may contain very high levels of chromium.

Copper is a member of the corrosion byproducts group. These are metals that are primarily present in tap water as a result of corrosion of pipes and plumbing fixtures. For these contaminants, action levels have been established instead of MCLs. When concentrations measured at the tap exceed the action level, specific steps must be taken to reduce the levels. While copper corrosion is a nuisance, it is rarely a human health concern.

Lead can be a major concern in water systems subject to high levels of corrosion. Because lead readily precipitates as carbonates and hydroxides, it is rarely present at high concentrations in natural waters. Nevertheless, elevated lead levels can be found in tap water if the water is corrosive and there is contact with lead in the pipes. Possible sources of lead include lead service connections, lead-tin solder, and lead-containing brass fixtures. As with all corrosion byproducts, concentrations of lead are highest at the tap during the "first flush" of water in the morning.

Mercury exists in three oxidation states (0, +I, +II). Metallic mercury readily volatilizes and can be transported through the atmosphere. Under reducing conditions, certain microorganisms, such as the sulfate-reducing bacteria, can methylate mercury, which further aids its transport and elevates its toxicity (Gilmour & Henry,

1991). Mercury is characterized by its ability to form strong complexes and concentrate through the food chain.

Thallium is most commonly found in the monovalent form, and it tends to be bound to soil particles. It is normally present in very low concentrations, below 1 μg/L in tap water (U.S. EPA, 1988). When higher concentrations are found, it is usually due to contamination from mining operations or metal manufacturing wastewaters.

The major **radionuclides** of concern in drinking waters are radium 226 and 228, uranium, and radon 222. Because it is a gas, radon poses a special concern. It is found in many groundwaters, depending on the local geology, and its concentrations can be highly variable. Based on data from the National Inorganic and Radionuclides Survey, 17 million Americans are expected to have radon levels in their drinking water in excess of the MCL (300 pCi/L), and 2.7 million are thought to have water above 1000 pCi/L.

Other **nonmetallic inorganics** with U.S. drinking water standards established or under development include selenium and arsenic and several nitrogen, sulfur, and halogen species. *Selenium* is toxic at high doses, yet it is also an essential nutrient. For this reason, its regulation is quite complex. From a chemical standpoint, selenium behaves much like sulfur. It has four oxidation states (–II, 0, +IV, +VI), and the two highest oxidation states tend to form the more soluble compounds. Both of these forms (selenites and selenates) are common in aerated waters. Selenium is relatively bioactive and can be methylated by microorganisms such as cyanobacteria (Bender et al., 1991). *Arsenic* has some similarities with selenium, but its toxicology is, if anything, more complex and the arsenic MCL is currently under review (U.S. EPA, 1996).

Among the inorganic *halogen species* of interest are *fluoride* (MCL also under review), which is added to many water supplies to prevent dental caries, some disinfection byproducts (*chlorite, chlorate,* and *bromate*), and *chlorine, chloramines,* and *chlorine dioxide,* which are all used as disinfectants in water treatment. A disinfectant residual (ca. 0.5 to 1.0 mg/L) is purposely added to finished waters in many systems.

2.3.2 ORGANICS

Organic chemicals may occur in drinking water as products of the disinfection process, as chemicals used in the water treatment process (e.g., acrylamide, epichlorohydrin), or as contaminants already present from other sources (e.g., solvents, pesticides). About two thirds of the standards established or proposed for chemicals in U.S. drinking waters are for organic species (see Table 2.1).

2.3.2.1 Disinfection Byproducts

The twentieth century has witnessed the rapid decline of many waterborne diseases (typhoid, cholera, amoebic dysentery, bacterial gastroenteritis) as a result of disinfection of water supplies. The proceedings of an ILSI conference, "Safety of Water Disinfection: Balancing Chemical and Microbial Risks" (Craun, 1993), offers a recent perspective on the important role of disinfection in the treatment of drinking water in developed and developing countries.

Direct chlorination was the first and is still among the most common disinfection methods in use worldwide. In the U.S. about 70% of the drinking water treatment systems serving a population >10,000 use chlorination for disinfection. Other disinfection processes used in major U.S. water supplies include, in decreasing order of frequency, treatment with chloramine, chlorine dioxide, or ozone (Farland & Gibb, 1993). Ozonation is more common in Europe (Singer, 1993).

The following paragraphs present a very brief overview of some of the chemicals and classes of chemicals identified as organic byproducts of the four major disinfection processes. (For more extensive reviews of this topic, see National Academy of Sciences, 1980; Singer, 1993; Richardson, 1998.) However, the distinction among the processes is not clear-cut, both because the same or similar byproducts are produced by different disinfectants and because of the tendency to use multiple disinfectants in series in a water treatment system. For example, chlorine or chloramine may be used to maintain a residual in the distribution system after disinfection with ozone. Thus, the differences in byproducts by process implied in the following paragraphs are in some cases more quantitative than qualitative.

Chlorine reacts with the naturally occurring organic matter (humic substances) during drinking water treatment to form a diverse group of organic byproducts. Drinking water standards have focused mainly on those haloorganics that have shown mutagenic or carcinogenic potential (Bull, 1982; Kringstad et al., 1983; Bull & Kopfler, 1991). Although hundreds of haloorganics have been identified in chlorinated waters, only a few are major contributors to the total organic halogen (TOX) in such waters (Table 2.2).

The *trihalomethanes (THMs)* are the best known and generally the most prevalent of the chlorination byproducts. They are also very easily measured and were discovered to be ubiquitous in chlorinated drinking waters in the U.S. in the mid-1970s (Symons et al., 1975). In most waters, chloroform ($CHCl_3$) is the main THM present, but depending on the bromide content of the water, bromodichloromethane ($CHBrCl_2$), dibromochloromethane ($CHBr_2Cl$), and bromoform ($CHBr_3$) are also found in substantial amounts. Total THM content of chlorinated drinking water is typically in the range of 10 to 100 μg/L.

The *haloacetic acids*, principally dichloro- and trichloroacetic acid, are the second most prevalent class of haloorganics found in chlorinated drinking waters. The total haloacetic acid level is typically in the same range as total THM (Reckhow & Singer, 1990; Singer, 1993).

Haloaldehydes, mainly trichloroacetaldehyde (which exists in water as its hydrate, chloral hydrate), and *haloketones* are also commonly detected in chlorinated waters at <10 μg/L (Bull & Kopfler, 1991). *Haloacetonitriles*, especially the dihaloacetonitriles, are typically found at total concentrations of 1 to 10μg/L and are considered to derive from chlorination of amino acids in proteinaceous material (Trehy & Bieber, 1981; de Leer et al., 1985; Bull & Kopfler, 1991). Two other nitrogen-containing byproducts of chlorination are *cyanogen chloride* (ca. 1 μg/L) and *chloropicrin* (trichloronitromethane; <1 μg/L).

An extensive list of C_3 to C_{10} haloacids have been isolated as chlorination byproducts. *Dichlorosuccinic acid* may be the most prevalent halogenated byproduct,

TABLE 2.2
Chemical Byproducts of Chlorination

Byproduct Class	Examples	Refs
Trihalomethanes	Chloroform	many
Haloacids (saturated)	Dichloroacetic acid	many
	Trichloroacetic acid	
Haloacids (unsaturated)	3,3-Dichloropropenoic acid	14
Halodiacids	2,2-Dichlorobutanedioic acid	14,15
	2,3-Dichlorobutenedioic acid	
Halohydroxyacids	3,3,3-Trichloro-2-hydroxypropanoic acid	14
Haloketones	1,1,1-Trichloropropanone	1,2,3
	1,1,1-Trichloro-2-butanone, Pentachloro-3-buten-2-one	4
Haloaldehydes	Chloral	16
	Dichloropropanal	4
	2,3,3-Trichloropropenal	
Haloketoacids	2,3-Dichloro-4-oxopentenoic acid	14
	2,5-Dichloro-4-bromo-3-oxopentanoic acid	
Haloacetonitriles	Dichloroacetonitrile	1,2,5
Haloalkanes	1,2-Dibromoethane, 1,2-Dibromopropane	13,14
Halophenols	2,4-Dichlorophenol	1,6
Chloroaromatic Acids	5-Chloro-2-methoxybenzoic acid	7,8
	Dichloromethoxybenzoic acid	
Halothiophenes	Tetrachlorothiophene	1
Chlorinated PAHs		9
MX and related compounds	MX, EMX, red-MX, ox-EMX	11,12
	2,2,4-Trichlorocyclopentene-1,3-dione	10
Misc. N-containing Haloorganics	Chloropicrin, Cyanogen chloride	16
Aliphatic Diacids (not halogenated)	Butenedioic acid	7,14,15
Aromatic Acids (not halogenated)	Benzoic acid	7,14,15

References:

1. Coleman et al., 1984
2. Reckhow & Singer, 1984
3. Sato et al., 1985
4. Kopfler et al., 1985
5. Oliver, 1983
6. Kringstad et al., 1985
7. de Leer et al., 1985
8. Seeger et al., 1985
9. Shiraishi et al., 1985
10. Smeds et al., 1995
11. Kronberg & Franzén, 1993
12. Strömberg et al., 1987
13. Fielding & Horth, 1986
14. Peters et al., 1994
15. Christman et al., 1980
16. Krasner et al., 1989

containing more than three carbon atoms (Christman et al., 1980). Among the unsaturated chlorinated acids, *MX* (which at acid pH exists mainly as its lactone, 3-chloro-4-(dichloromethyl)-5-hydroxy-2(5H)-furanone) is a potent mutagen and animal carcinogen that is frequently detected at low levels (<0.01 µg/L) in chlorinated waters (Komulainen et al., 1997; Melnick et al., 1997).

Despite the large number of halogenated chlorination byproducts identified in the past 20 years, only about 50% of the TOX is accounted for by these byproducts (Singer, 1993). It is likely that the remainder consists of higher-molecular-weight compounds not amenable to analysis and trace quantities of other haloorganics.

Although regulatory attention has focused on the halogenated byproducts, most of the organic byproducts of chlorination probably do not contain chlorine or bromine (Christman et al., 1983). These include aliphatic mono- and dibasic acids, aromatic carboxylic acids, and a number of aldehydes and aldo- and ketoacids that are also common products of ozonation.

Chloramine treatment has been developed as an alternative to chlorination and can result, with proper controls, in reduced levels of THMs as byproducts (Singer, 1993). Chloramines are formed by the reaction of free chlorine with ammonia. The reaction is stepwise, giving monochloramine (NH_2Cl) followed by dichloramine ($NHCl_2$). The latter is unstable, forming nitrogen gas and some nitrate, so in chloramination, as the chlorine:ammonia ratio is increased beyond 1.0, the "combined chlorine residual" available for disinfection decreases until a "breakpoint" (minimum) is reached when ammonia and monochloramine have been consumed.

A wide range of compounds have been identified from the reaction of monochloramine with organics (Table 2.3). Monochloramine reacts with amines and amino acids to form N-chloroamines (Crochet & Kovacic, 1973), a reaction that can also occur with free chlorine. It also can add to activated carbon-carbon double bonds to give β-chloroamines or chlorohydrins (Minisci & Galli, 1965). Chlorine substitution on activated aromatic rings (e.g., converting phenols to chlorophenols) has also been observed (Burttschell et al., 1959).

Monochloramine is far less reactive with most organic compounds than free chlorine. Significant chloramination byproducts include cyanogen chloride, chloropicrin, dichloroacetic acid, and much smaller amounts of chloroform and trichloroacetic acid (which may be formed due to the presence of a small free chlorine residual). MX has been reported to occur in the chloramination of humic materials, although the amount measured was less than 25% of that formed during chlorination (Backlund et al., 1988). Johnson and Jensen (1986) found that at high monochloramine/carbon molar ratios (ca. 10 mg/mg) about 5% of the oxidant demand was converted to TOX. Identical results were found for free chlorine at comparable chlorine/carbon ratios. However, most of the haloorganic chloramination byproducts are nonvolatile, higher molecular-weight compounds, as chloramine is a much weaker oxidant than chlorine and is much less likely to oxidatively fragment humic molecules.

Chlorine dioxide (ClO_2), used as a water disinfectant, is generated by the reaction of chlorite (ClO_2^-) with chlorine. As with chloramination, the advantage of chlorine dioxide treatment over chlorination is that there is less chlorine substitution relative to oxidation. However, the inorganic species formed (chlorine dioxide, chlorite, chlorate) are also potentially toxic.

Most of the products of reaction of chlorine dioxide with organic matter are oxygenated species (Table 2.4). Although haloacids have been observed in laboratory studies, trihalomethanes have not been detected. Like the other chlorine-containing oxidants,

TABLE 2.3
Chemical Byproducts of Chloramines

Byproduct Class	Examples	Refs
Cyanogen Halides	Cyanogen chloride, Cyanogen bromide	many
Chloroacids	Dichloroacetic acid	1
	Trichloroacetic acid	
	6,6-Dichlorohexanoic acid	
Unsaturated Chloroacids	3,3,-Dichloropropenoic acid	1
Chlorohydroxy Acids	2-Chloro-4-hydroxybutanoic acid	1
	2,3-Dichloro-3,3-dihydroxypropanoic acid	
Unsaturated Chlorohydroxy Acids	4-Chloro-4-hydroxypentenoic acid	1
Chlorinated Ketones and Aldehydes	Chloroacetaldehyde	1
	Dichloroacetaldehyde	
	Chloropropanal	
	3-Chlorobutanal	
	3,3-Dichloropropanal	
	1,6,6-Trichlorohexanal	
	Chloropropanone	
Chlorodiacids	Dichlorosuccinic acid	1
Nitriles		2,3
Aldehydes		3
C-Chloroamines		4,9
N-Chloroamines		5,7,8
Chlorohydrins		6,9
MX and Derivatives	MX, EMX, red-MX, ox-EMX	1,10
Epoxides		6

References:

1. Kanniganti, 1990, and Kanniganti et al., 1992
2. Hausler & Hausler, 1930
3. Le Cloirec & Martin, 1985
4. Neale, 1964
5. Scully, 1986
6. Carlson & Caple, 1977
7. Jensen & Johnson, 1989
8. Crochet & Kovacic, 1973
9. Minisci & Galli, 1965
10. Backlund et al., 1988

chlorine addition/substitution products are favored at low oxidant-to-carbon ratios, and oxidation reactions are favored at high ratios.

Ozone (O_3) has been used extensively as a water disinfectant in Europe but not so much in the United States. Among the products reported from the ozonation of drinking waters are simple aldehydes, aliphatic and aromatic acids, some ketoacids, and other polycarbonyl products (see Table 2.5). Many of these products are also produced by the chlorine-containing disinfectants but sometimes at much lower levels. For example, total aldehydes in ozonated drinking water appear to range from a few μg/L to several hundred μg/L, depending on how much organic matter is present in the water and the applied ozone/organic carbon ratio (Singer, 1993). The tendency of ozone to form organic free radicals results in a large amount of uncharacterized

TABLE 2.4
Chemical Byproducts of Chlorine Dioxide

Byproduct Class	Examples	Refs
Oxychlorine Cmpds. (Inorg.)	Chlorite, Chlorate	7,8
Aliphatic Monoacids	Acetic acid, Butanoic acid, Pentanoic acid, Hexanoic acid, 2-Ethylhexanoic acid	1, 2,11
Aliphatic Diacids	Oxalic acid	5
	Succinic acid, Glutaric acid, Adipic acid	2
Haloacids	Dichloroacetic acid, Chloromalonic acid, Chlorosuccinic acid, 9-Chloro-10-methylstearate	2, 10
Aliphatic Aldehydes	Acetaldehyde, Propanal	3
Halogenated Ketones	1,1,3,3-Tetrachloropropanone	1
Unsaturated Ketones	2,3,4-Trimethylcyclopent-2-en-1-one	1
	2,6,6-Trimethyl-2-cyclohexene-1,4-dione	
Maleic Acid &	Maleic acid	5
Derivatives	2-*tert*-Butylmaleic acid,	1
	2-Ethyl-3-methylmaleic acid	
Styrenes	3-Ethylstyrene, 4-Ethylstyrene	1
Aromatics	Benzoic acid	1
	p-Benzoquinone, Hydroxy-PAHs	6,9
Naphthalene Derivatives	Naphthalene	1
	1-Methylnaphthalene	
Furan derivatives	Methylfurancarboxylic acid	2
Epoxides		4,12

References:

1. Richardson et al., 1994
2. Colclough, 1981
3. Stevens et al., 1978
4. Legube et al., 1981
5. Masschelein, 1979
6. Wajon et al., 1982

7. Steinbergs, 1986
8. Werdehoff & Singer, 1987
9. Liukkonen et al., 1983
10. Ghanbari et al., 1983
11. Somsen, 1960
12. Lindgren & Svahn, 1966

higher-molecular-weight material as byproducts of ozonation. Also, the peroxides and hydroperoxides are so reactive that they do not persist and are difficult to identify.

In waters containing significant levels of bromide, ozone readily oxidizes bromide to hypobromous acid (HOBr), which reacts with organic matter to produce brominated analogues of the chlorinated byproducts formed during chlorination (Glaze et al., 1993). These include bromoform, brominated acetic acids and acetonitriles, bromopicrin, and cyanogen bromide. Inorganic bromate is another significant byproduct under these conditions.

Finally, it should be noted that when the primary disinfection process is other than chlorination, and especially when it is ozonation, a secondary disinfectant (e.g., chlorine residual) is often added to maintain control of pathogens in the distribution system. Many of the primary oxidation byproducts can then react further to form

TABLE 2.5
Chemical Byproducts of Ozone

Byproduct Class	Examples	Refs
Oxyhalogen Cmpds. (Inorg.)	Bromate, Hypobromite	11,12,13
Organic Bromides	Dibromoacetic acid, Bromoform, Bromopicrin	9,10,12
Inorganic Peroxides	Hydrogen peroxide	many
Aliphatic Aldehydes	Formaldehyde, Acetaldehyde, Propanal	many
Aromatic Aldehydes	Benzaldehyde, Ethylbenzaldehyde	7,17
Aliphatic Ketones	Acetone, 3-Hexanone	1, 14
Aromatic Ketones	Acetophenone, 4-Phenyl-2-butanone	7
Aliphatic Monoacids	Hexanoic acid	16
	Acetic acid	3,4
Aliphatic Diacids	Oxalic acid, Malonic acid	4,8
	Succinic acid	7
Aliphatic Monoacid-Monocarbonyls	Glyoxalic Acid, Pyruvic Acid	many
	Oxobutanoic acid, 4-Oxopentanoic acid, 4-Oxo-2-butenoic acid	1,7
Aliphatic Diacid-Monocarbonyls	Ketomalonic acid, Ketosuccinic acid, Ketoglutaric acid	1
Aliphatic Dicarbonyls	Glyoxal, Methylglyoxal	many
	1,2-Dioxobutane, Dioxopentane	1
Aliphatic Monoacid-Dicarbonyls	Dioxopropanoic acid, Dioxobutanoic acid, Dioxopentanoic acid	1
Aliphatic Hydroxyacids	Hydroxymalonic acid	4
Aromatic Acids	Benzoic acid, 3,5-Dimethylbenzoic acid	4,5,6,7
Epoxides		2,15
Miscellaneous	5-Methoxy-α-pyrone	4

References:

1. Le Lacheur et al., 1993
2. Carlson & Caple, 1977
3. Lawrence, 1977
4. Benga, 1980
5. Paramisigamani et al., 1983
6. Killips et al., 1985
7. Glaze, 1986
8. Edwards, 1990
9. Cooper et al., 1986
10. Daniel et al., 1989
11. Haag & Hoigné, 1983
12. Glaze et al., 1993
13. Krasner et al., 1993
14. Fawell & Watts, 1984
15. Chen et al., 1979
16. Chappell et al., 1981
17. Lawrence et al., 1980

secondary byproducts; for example, acetaldehyde can be converted to chloral (McKnight & Reckhow, 1992).

2.3.2.2 Volatile Organic Compounds (VOCs)

In addition to the disinfection byproducts, many of which have significant volatility, there is another group of water contaminants that have been loosely referred to as volatile organic compounds, or VOCs (see Table 2.1). These include a series of 1-, 2-, and 3-carbon chlorinated compounds, several chlorobenzenes, hexachlorocyclo-pentadiene, and some common aromatic hydrocarbons. Although a few of the chlorinated compounds may be formed in small amounts as disinfection byproducts, these VOCs are mainly high-volume industrial solvents and intermediates, and drinking water standards have been established or proposed because of their widespread commercial applications and their relative persistence and mobility.

Of the **1-, 2-, and 3-carbon chlorinated compounds**, several are solvents (methylene chloride, carbon tetrachloride, trichloroethylene, tetrachloroethylene, 1,1,1-trichloroethane, 1,2- dichloropropane), some are intermediates (vinyl chloride, 1,1-dichloroethylene, 1,2-dichloroethane, 1,1,2-trichloroethane), and others are mainly byproducts of industrial processes (*cis*- and *trans*-1,2-dichloroethylene). The **chlorobenzenes** (chlorobenzene, *o*- and *p*-dichlorobenzene, 1,2,4-trichlorobenzene, hexachlorobenzene) have found use as solvents, intermediates, and pesticides. **Hexachlorocyclopentadiene** has been used principally as an intermediate in the manufacture of a number of chlorinated pesticides. The **aromatic hydrocarbons** include four of the most common aromatic solvents (benzene, toluene, ethylbenzene, xylenes) and styrene, an important monomer.

2.3.2.3 Pesticides

In the late 1970s reports began to appear of pesticides detected in drinking water supplies, and the Safe Drinking Water Act Amendments of 1986 specifically designated a number of pesticides for regulatory action. To date, EPA has set MCLs and MCLGs for about 25 pesticides (see Table 2.1) and has developed Drinking Water Health Advisories for many others (U.S. EPA, 1996). Standards have been set for **pesticides** from several different chemical classes, such as the organochlorines (chlordane, heptachlor, lindane), carbamates (aldicarb, carbofuran), triazines (atrazine, simazine), phenoxy herbicides (2,4-D, 2,4,5-TP) and pentachlorophenol.

2.3.2.4 Miscellaneous

Several other classes of potential drinking water contaminants have received some attention (Table 2.1). Control of **chemicals used in water treatment** (acrylamide, epichlorohydrin) is focused on treatment technique rather than standards. Standards have been set for two **plasticizers**, di(2-ethylhexyl)adipate (DEHA) and di(2-ethylhexyl)phthalate (DEHP), and for **polychlorinated biphenyls (PCBs)** as a class. Although polychlorinated dibenzo-*p*-dioxins (PCDDs) and polychlorinated dibenzofurans (PCDFs) are widely distributed in the environment, they exhibit very low solubility in water and are detected more commonly in sediments (IARC, 1997), and only one of the congeners, **2,3,7,8-tetrachlorodibenzo-*p*-dioxin (TCDD)**, has a drinking water standard. Similarly, the polycyclic aromatic hydrocarbons (PAHs) are ubiquitous byproducts of the combustion of fossil fuels, wood, tobacco, and other sources, but they are relatively insoluble in water and, at present, only **benzo[*a*]pyrene** has an MCL and MCLG.

REFERENCES

Andreae, M. O. (1983) The Determination of the Chemical Species of Some of the Hydride Elements in Seawater: Methodology and Results. NATO Conference 4:1–19.

AWWA (1986) 1984 Water Quality Operating Data. American Water Works Association, Denver, CO.

AWWA (1987) 1985 Water Quality Operating Data. American Water Works Association, Denver, CO.

Backlund, P., Kronberg, L., and Tikkanen, L. (1988) Formation of Ames Mutagenicity and of the Strong Bacterial Mutagen 3-chloro-4-(dichloromethyl)-5-hydroxy-2-(5H)-furanone and other Halogenated Compounds During Disinfection of Humic Water. Chemosphere 17:1329–1336.

Bender, J., Gould, J. P., and Vatcharapijarn, Y. (1991) Uptake, Transformation and Fixation of Selenium (VI) by a Mixed Selenium-tolerant Ecosystem. Water Air Soil Polution 59:359–368.

Benga, J. (1980) Oxidation of Humic Acid by Ozone or Chlorine Dioxide [dissertation]. Miami University (Ohio).

Brass, H. J., Feige, M. A., Halloran, T., et al. (1977) The National Organic Monitoring Survey: Samplings and Analyses for Purgeable Organic Compounds. In Pojasek RB (Editor). Drinking Water Quality Enhancement Through Source Protection. Ann Arbor Science Publ., Inc., Ann Arbor, MI, pp 393–416.

Bull, R. J. (1982) Toxicological Problems Associated with Alternative Methods of Disinfection. Journal American Water Works Association 74:642-648.

Bull, R. J., and Kopfler, F. C. (1991) Health Effects of Disinfectants and Disinfection Byproducts. AWWA Research Foundation, Denver, CO.

Burttschell, R.H., Rosen, A. A., Middleton, F. M., et al. (1959). Journal of the American Water Works Association 51:205.

Carlson, R., and Caple, R. (1977) Chemical Biological Implications of Using Chlorine and Ozone for Disinfection. U.S. Environmental Protection Agency, Washington Report No.: EPA-600/3-77-066 88p. Ecological Research Series.

Chen, P. N., Junk, G. A., and Svec, H. J. (1979) Reactions of Organic Pollutants. I. Ozonation of Acenaphthylene and Acenaphthene. Environmental Science and Technology 13:451–454.

Christman, R. F., Johnson, J. D., Pfaender, F. K., et al. (1980) Chemical Identification of Aquatic Humic Chlorination Products. In Jolley Robert L., Brungs William A., Cumming Robert B. (Editors). Water Chlorination: Environmental Impact and Health Effects. Volume 3. pp 75–83.

Christman, R. F., Norwood, D. L., Millington, D. S., Johnson, J. D., and Stevens, A. A. (1983) Identity and Yields of Major Halogenated Products of Aquatic Fulvic Acid Chlorination. Environmental Science and Technology 17:625-628.

Colclough, C. A. (1981). Organic Reaction Products of Chlorine Dioxide and Natural Aquatic Fulvic Acids. Unpublished masters thesis, University of North Carolina at Chapel Hill.

Coleman, W. E., Munch, J. W., Kaylor, W. H., et al. (1984) Gas Chromatography/Mass Spectroscopy Analysis of Mutagenic Extracts of Aqueous Chlorinated Humic Acid. A Comparison of the Byproducts to Drinking Water Contaminants. Environmental Science and Technology 18:674–681.

Cooper, W. J., Amy, G. L., Moore, C. A., et al. (1986) Bromoform Formation in Ozonated Groundwater Containing Bromide and Humic Substances. Ozone: Science & Engineering 8:63–76.

Craun, G. F. (Editor) (1993) Safety of Water Disinfection: Balancing Chemical and Microbial Risks. ILSI Press, Washington, D.C.

Crochet, R. A., and Kovacic, P. (1973) Conversion of o-Hydroxyaldehydes and Ketones into o-Hydroxyanilides by Monochloramine. Journal of the Chemical Society. Chemical Communications 1973:716–717.

Croué, J.-P., and Reckhow, D. A. (1989) The Destruction of Chlorination Byproducts with Sulfite. Environmental Science and Technology 23:1412–1419.

Daniel, P. A., Meyerhofer, P. F., Lanier, M., et al. (1989) Impact of Ozonation on Formation of Brominated Organics. In Bollyky, L. J., (Editor), Ozone in Water Treatment; Proceedings of the 9th Ozone World Congress: International Ozone Association, Zurich pp 348–360.

de Leer, E. W. B., Damste, J. S. S., Erkelens, C., et al. (1985) Identification of Intermediates Leading to Chloroform and C-4 Diacids in the Chlorination of Humic Acid. Environmental Science and Technology 19:512–522.

Eckel, W. P., and Jacob, T. A. (1988) Ambient Levels of 24 Dissolved Metals in U.S. Surface and Ground Waters. In Proceedings of the 196th Meeting of the American Chemcial Society, Division of Environmental Chemistry pp 371–372.

Edwards, M. (1990) Transformation of Natural Organic Matter, Effect of Organic Matter-Coagulation Interactions, and Ozone-Induced Particle Destabilization [dissertation]. University of Washington at Seattle.

Farland, W. H., and Gibb, H. J. (1993) U.S. Perspective on Balancing Chemical and Microbial Risks of Disinfection. In Craun, G. F. (Editor). Safety of Water Disinfection: Balancing Chemical and Microbial Risks. ILSI Press, Washington, D.C., pp3–10.

Fawell, J. K., and Watts, C. D. (1984) The Nature and Significance of Organic By-products of Ozonation — A Review. In Seminar on Ozone in U.K. Water Treatment: The Institution of Water Engineers and Scientists, London.

Fielding, M., and Horth, H. (1986). Water Supply 4:103–126.

Ghanbari, H. A., Wheeler, W. B., and Kirk, J. R. (1983) Reactions of Chlorine and Chlorine Dioxide With Free Fatty Acids, Fatty Acid Esters, and Triglycerides. In Water Chlorination: Environmental Impact and Health Effects. Volume 4, Book 1, Chemistry and Water Treatment. Proceedings of the Fourth Conference on Water Chlorination. Robert L. Jolley, William A. Brungs, Joseph A. Cotruvo, Robert B. Cumming, Jack S. Mattice, and Vivian A. Jacobs, (Editors). Pacific Grove, CA pp 167–177.

Gilmour, C. C., and Henry, E. A. (1991) Mercury Methylation in Aquatic Systems Affected by Acid Deposition. Environmental Pollution 71:131–169.

Glaze, W. H. (1986) Reaction Products of Ozone: A Review. Environmental Health Perspectives 69:151–157.

Glaze, W. H., Weinberg, H. S., and Cavanagh, J. E. (1993) Evaluating the Formation of Brominated DBPs During Ozonation. Journal of the American Water Works Association 85:96–103.

Greathouse, D. G., and Craun, G. F. (1978) Cardiovascular Disease Study — Occurrence of Inorganics in Household Tap Water and Relationships to Cardiovascular Mortality Rates. In Hemphil, D. D. (Editor). Trace Substances in Environmental Health. University of Missouri, pp 31–39.

Haag, W. R., and Hoigné, J. (1983) Ozonation of Bromide-Containing Water: Kinetics of Formation of Hypobromous Acid and Bromate. Environmental Science and Technology 17:261–267.

Hansch, C., Leo, A., and Hoekman, D. (1995). Exploring QSAR — Hydrophobic, Electronic and Steric Constants. American Chemical Society, Washington.

Hausler, C. R., and Hausler, M. L. (1930) Research on Chloramines. I. Orthochlorobenzal-chlorimine and Anisalchlorimine. Journal of the American Chemical Society 52:2050–2054.

IARC (1990) IARC Monographs on the Evaluation of Carcinogenic Risks to Humans (Vol. 49): Chromium, Nickel and Welding. International Agency for Research on Cancer, Lyon, pp 49–256.

IARC (1997) IARC Monographs on the Evaluation of Carcinogenic Risks to Humans (Vol. 69): Polychlorinated Dibenzo-*para*-dioxins and Polychlorinated Dibenzofurans. International Agency for Research on Cancer, Lyon.

Jensen, J. N., and Johnson, J. D. (1989) Specificity of the DPD and Amperometric Titration Methods for Free Available Chlorine: A Review. Journal of the American Water Works Association 81:59–64.

Johnson, J. D., and Jensen, J. N. (1986) THM and TOX Formation: Routes, Rates, and Precursors. Journal American Water Works Association 78:156-162.

Kanniganti, R. (1990). Characterization and Gas Chromatography/Mass Spectrometry Analysis of Mutagenic Extracts of Aqueous Monochloraminated Fulvic Acid. Unpublished masters thesis, University of North Carolina at Chapel Hill.

Kanniganti, R., Johnson, J. D., Ball, L. M., et al. (1992) Identification of Compounds in Mutagenic Extracts of Aqueous Monochloraminated Fulvic Acid. Environmental Science and Technology 26:1998–2004.

Komulainen, H., Kosma, V.-M., Vaittinen, S.-L., et al. (1997) Carcinogenicity of the Drinking Water Mutagen 3-Chloro-4-(dichloromethyl)-5-hydroxy-2(5H)-furanone in the Rat. Journal of the National Cancer Institute 89:848–856.

Kopfler, F. C., Ringhand, H. P., Coleman, W. E., et al. (1985) Reactions of Chlorine in Drinking Water, with Humic Acids and *In Vivo*. In Jolley, R. L., Bull, R. J., David, W. P., et al. (Editors) Water Chlorination: Environmental Impact and Health Effects: Lewis Publishers, Chelsea, MI pp 161–173.

Krasner, S. W., Glaze, W. H., Weinberg, H. S., et al. (1993) Formation and Control of Bromate During Ozonation of Waters Containing Bromide. Journal of the American Water Works Association 85:73–81.

Krasner, S. W., McGuire, M. J., Jacangelo, J. G., et al. (1989) The Occurrence of Disinfection Byproducts in U.S. Drinking Water. Journal of the American Water Works Association 81:41–53.

Kringstad, K. P., Ljungquist, P. O., de Sousa, F., et al. (1983) Stability of 2-Chloropropenal and Some Other Mutagens Formed in the Chlorination of Softwood Kraft Pulp. Environmental Science and Technology 17:468–471.

Kringstad, K. P., de Sousa, F., and Strömberg, L. M. (1985) Studies on the Chlorination of Chlorolignins and Humic Acid. Environmental Science and Technology 19:427–431.

Kronberg, L., and Franzén, R. (1993) Determination of Chlorinated Furanones, Hydroxyfuranones, and Butenedioic Acids in Chlorine-treated Water and in Pulp Bleaching Liquor. Environmental Science and Technology 27:1811–1818.

Lange's Handbook of Chemistry (1992) Dear, John A (Editor). McGraw-Hill Publishers, New York.

Lawrence, J. (1977) The Oxidation of Some Haloform Precursors with Ozone. In Proceedings of the 3rd Ozone World Congress: International Ozone Institute.

Lawrence, J., Tosine, H., Onuska, F. I., et al. (1980) The Ozonation of Natural Waters: Product Identification. Ozone: Science and Engineering 2:55–64.

Le Cloirec, C., and Martin, G. (1985) Evolution of Amino Acids in Water Treatment Plants and the Effect of Chlorination on Amino Acids. In Jolley, R. L., Bull, R. J., David, W. P., et al. (Editors). Water Chlorination: Environmental Impact and Health Effects: Lewis Publishers, Chelsea, MI pp 821–834.

Le Lacheur, R. M., Sonnenberg, L. B., Singer, P. C., et al. (1993) Identification of Carbonyl Compounds in Environmental Samples. Environmental Science and Technology 27:2745–2753.

Legube, B., Langlais, B., Sohm, B., et al. (1981) Identification of Ozonation Products of Aromatic Hydrocarbon Micropollutants: Effect on Chlorination and Biological Filtration. Ozone: Science and Engineering 3:33–48.

Leo, A., Hansch, C., and Elkins, D. (1971). Partition Coefficients and their Uses. Chemical Reviews 71:525.

Lide, D. R (Editor) (1995) Handbook of Chemistry and Physics, 76th Ed. CRC Press, Boca Raton, FL.

Lindgren, B. O., and Svahn, C. M. (1966) Reactions of Chlorine Dioxide with Unsaturated Compounds. II. Methyl Oleate. Acta Chemica Scandinavia 20:211–218.

Liukkonen, R. J., Lin, S., Oyler, A. R., et al. (1983) Product Distribution and Relative Rates of Reaction of Aqueous Chlorine and Chlorine Dioxide With Polynuclear Aromatic Hydrocarbons. In Jolley, R. L., Brungs, W. A., Cotruvo, J. A., Cumming, R. B., Mattice, J. S., and Jacobs, V. A., eds., Water Chlorination: Environmental Impact and Health Effects. Volume 4, Book 1, Chemistry and Water Treatment. Proceedings of the Fourth Conference on Water Chlorination. Pacific Grove, CA pp 151–165.

Lyman, W. J., Reehl, W. F., and Rosenblatt, D. H. (1990). Handbook of Chemical Property Estimation Methods. American Chemical Society, Washington, D.C.

Mabey, W., and Mill, T. (1978). Journal of Physical and Chemical Reference Data 7:383.

Mackay, D., and Shiu, W. Y. (1981) Critical Review of Henry's Law Constants for Chemicals of Environmental Interest. Journal of Physical and Chemical Reference Data 10:1175–1199.

Mackay, D., Shiu, W. Y., and Ma, K. C. (1992a). Illustrated Handbook of Physical-Chemical Properties and Environmental Fate for Organic Chemicals. Volume 1. Monoaromatic Hydrocarbons, Chlorobenzenes, and PCBs. Lewis Publishers, Chelsea, MI.

Mackay, D., Shiu, W. Y., and Ma, K. C. (1992b). Illustrated Handbook of Physical-Chemical Properties and Environmental Fate for Organic Chemicals. Volume II. Polynuclear Aromatic Hydrocarbons, Polychlorinated Dioxins, and Dibenzofurans. Lewis Publishers, Chelsea, MI.

Mackay, D., Shiu, W. Y., and Ma, K. C. (1993). Illustrated Handbook of Physical-Chemical Properties and Environmental Fate for Organic Chemicals. Volume III. Volatile Organic Chemicals. Lewis Publishers, Chelsea, MI.

Mackay, D., Shiu, W. Y., and Ma, K. C. (1995). Illustrated Handbook of Physical-Chemical Properties and Environmental Fate for Organic Chemicals. Volume IV. Oxygen, Nitrogen, and Sulfur-containing Compounds. Lewis Publishers, Chelsea, MI.

Masschelein, W. J. (1979). Chlorine Dioxide: Chemistry and Environmental Impact of Oxychlorine Compounds. Ann Arbor Science Publishers, Ann Arbor, MI.

McKnight, A., and Reckhow, D. A. (1992) Reactions of Ozonation Byproducts With Chlorine and Chloramines. In 1992 Annual Conference Proceedings; American Water Works Association, pp 399–409.

Melnick, R. L., Boorman, G. A., and Dellarco, V. (1997) Water Chlorination, 3-Chloro-4-(dichloromethyl)-5-hydroxy-2(5H)-furanone (MX), and Potential Cancer Risk. Journal of the National Cancer Institute 89:832–833.

Minisci, F., and Galli, R. (1965). A New, Highly Selective Type of Aromatic Substitution. Homolytic Amination of Phenolic Ethers. Tetrahedron Letters 1965:433-436.

Morris, J. C. (1967) Kinetics of Reactions between Aqueous Chlorine and Nitrogenous Compounds. In Faust, S. D., and Hunter, J. V., (Editors). Principles and Applications of Water Chemistry. John Wiley and Sons, New York, pp 28–53.

National Academy of Sciences (1980) Drinking Water and Health: Vol. 2. National Academy Press, Washington, D.C.

Neale, R. (1964) The Chemistry of Ion Radicals: The Free-Radical Addition of N-Chloramines to Olefinic and Acetylenic Hydrocarbons. Journal of the American Chemical Society 86:5340–5342.

Ohanian, E. V. (1986) Health Effects of Corrosion Products in Drinking Water. Trace Substances and Environmental Health 20:122–138.

Oliver, B. G. (1983) Dihaloacetonitriles in Drinking Water: Algae and Fulvic Acid as Precursors. Evironmental Science and Technology 17:80–83.

Page, G. W. (1981) Comparison of Groundwater and Surface Water for Patterns and Levels of Contamination by Toxic Substances. Environmental Science and Technology 15:1475–1481.

Paramisigamani, V., Malaiyandi, M., Benoit, F. M., et al. (1983) Identification of Ozonated and/or Chlorinated Residues of Fulvic Acids. In Proceedings of the 6th Ozone World Congress: International Ozone Association, Vienna, VA p 88.

Pellizzari, E. D., Perritt, K., Hartwell, T. D., et al. (1987a) Total Exposure Assessment Methodology (TEAM) Study: Elizabeth and Bayonne, New Jersey; Devils Lake, North Dakota; and Greensboro, North Carolina. Vol. II. U.S. Environmental Protection Agency, Washington, D.C.

Pellizzari, E. D., Perritt, K., Hartwell, T. D., et al. (1987b) Total Exposure Assessment Methodology (TEAM) Study: Selected Communities in Northern and Southern California. Vol. III. U.S. Environmental Protection Agency, Washington, D.C.

Peters, R. J. B., de Leer, E. W. B., and Versteegh, J. F. M. (1994) Identification of Halogenated Compounds Produced by Chlorination of Humic Acid in the Presence of Bromide. Journal of Chromatography A 686:253–261.

Reckhow, D. A., and Singer, P. C. (1984) The Removal of Organic Halide Precursors by Preozonation and Alum Coagulation. Journal of the American Water Works Association 76:151–157.

Reckhow, D. A., and Singer, P. C. (1990) Chlorination By-products in Drinking Water: From Formation Potentials to Finished Water Concentrations. Journal of the American Water Works Association 82:173–180.

Richardson, S. (1998) Drinking Water Disinfection Byproducts. In Meyers, R. A. (Editor). Encyclopedia of Environmental Analysis and Remediation: Vol. III. John Wiley & Sons, New York, pp. 1398–1421.

Richardson, S. D., Thruston, A. D. Jr., Collette, T. W., et al. (1994) Multispectral Identification of Chlorine Dioxide Disinfection Byproducts in Drinking Water. Environmental Science and Technology 28:592–599.

Sato, T., Mukaida, M., Ose, Y., et al. (1985) Mutagenicity of Chlorinated Products from Soil Humic Substances. Sci Total Environ 46:229-242.

Schwarzenbach, R., Gschwend, P.M., and Imboden, D. M. (1993). Environmental Organic Chemistry. John Wiley & Sons, New York.

Scully, F. E. Jr. (1986) N-Chloro Compounds: Occurrence and Potential Interference in Residual Analysis. In Proceedings Water Quality Technology Conference; Advances in Water Analysis and Treatment, Houston, TX pp 611–622.

Seeger, D. R., Moore, L. A., and Stevens, A. A. (1985) Formation of Acidic Trace Organic Byproducts from Chlorination of Humic Acids. In Jolley, R. L., Bull, R. J., David, W. P., et al. (Editors). Water Chlorination: Environmental Impact and Health Effects: Lewis Publishers, Chelsea, MI pp 859–873.

Shacklette, H. T., and Boerngen, J. G. (1984) Element Concentration in Soils and Other Surficial Materials of the Conterminous United States. U.S. Geological Survey Professional Paper 1270.

Shank, R. C., and Whittaker, C. (1988) Formation of Enotixic Hydrazine by the Chloramination of Drinking Water. University of California Water Resources Center, California Report No.: Technical Completion Report, Project No. W-690.

Shiller, A. M., and Boyle, E. A. (1987) Variability of Dissolved Trace Metals in the Mississippi River. Goechim Cosmochim Acta 51:3273–3278.

Shiraishi, H., Polkington, N. H., Otsuke, A., and Fuwa, K. (1985) Occurrence of Chlorinated Polynuclear Aromatic Hydrocarbons in Tap Water. Environmental Science and Technology 19:585.

Singer, P. C. (1993) Formation and Characterization of Disinfection By-products. In Craun, G. F. (Editor). Safety of Water Disinfection: Balancing Chemical and Microbial Risks. ILSI Press, Washington, D.C., pp 201–219.

Smeds, A., Franzén, R., and Kronberg, L. (1995) Occurrence of Some Chlorinated Enol Lactones and Cyclopentene-1,3-diones in Chlorine-Treated Waters. Environmental Science and Technology 29:1839–1844.

Smith, R. M., and Martell, A. E. (1974) Critical Stability Constants. Plenum Press, New York.

Snoeyink, V. L., and Jenkins, D. (1980) Water Chemistry. John Wiley & Sons, New York, pp 392–399.

Somsen, R.A. (1960) Oxidation of Some Simple Organic Molecules with Aqueous Chlorine Dioxide Solutions. Journal of the Technical Association of the Pulp and Paper Industry 43:154-160.

Steinbergs, C. Z. (1986) Removal of By-products of Chlorine and Chlorine Dioxide at a Hemodialysis Center. Journal of the American Water Works Association 78:94–98.

Stevens, A. A., Slocum, C. J., Seeger, D. R., et al. (1978) Chlorination of Organics in Drinking Water. In Jolley, R. L. (Editor). Water Chlorination: Environmental Impact and Health Effects. Volume 1. Proceedings of the Conference on the Environmental Impact of Water Chlorination, Oak Ridge, TN pp 77–104.

Strömberg, L. M., de Sousa, F., Ljungquist, P., McKague, B., and Kringstad, K. P. (1987) An Abundant Chlorinated Furanone in the Spent Chlorination Liquor from Pulp Bleaching. Environmental Science and Technology 21:754-756.

Stumm, W. R., and Morgan, J. J. (1996) Aquatic Chemistry: Chemical Equilibria and Rates in Natural Waters. John Wiley & Sons, New York.

Symons, J. M., Bellar, T. A., Carswell, J. K., et al. (1975) National Organics Reconnaissance Survey for Halogenated Organics. Journal of the American Water Works Association 67: 634–647.

Trehy, M. L., and Bieber, T. I. (1981) Detection, Identification, and Quantitative Analysis of Dihaloacetonitriles in Chlorinated Natural Waters. In Keith, L. H. (Editor). Advances in the Identification and Analysis of Organic Pollutants in Water, Volume 2: Ann Arbor Science, Ann Arbor, MI pp 941–975.

US EPA (1987) Health Assessment Document for Beryllium. Environmental Criteria and Assessment Office, Office of Health and Environmental Assessment, U.S. Environmental Protection Agency, Washington, D.C. Report No.: EPA/600/8-84-026F.

US EPA (1988) Health and Environmental Effects Document for Thallium and Compounds. U.S. Environmental Protection Agency, Office of Solid Waste and Emergency Response, Cincinnati, OH Report No.: ECAO-CIN-G031.

US EPA (1990) National Survey of Pesticides in Drinking Water Wells, Phase 1 Report. U.S. Environmental Protection Agency, Washington, D.C.: Report No. EPA 570/9-90-015.

US EPA (1992) Another Look: National Survey of Pesticides in Drinking Water Wells, Phase 2 Report. U.S. Environmental Protection Agency, Washington, D.C.: Report No. EPA 570/9-91-020.

US EPA (1996) Drinking Water Regulations and Health Advisories. U.S. Environmental Protection Agency, Washington, D.C.: EPA 822-B-96-002.

Verschueren, K. (1996) Handbook of Environmental Data on Organic Chemicals, 3rd ed. Van Nostrand Reinhold, New York.

Wajon, J. E., Rosenblatt, D. H., and Burrows, E. P. (1982) Oxidation of Phenol and Hydroquinone by Chlorine Dioxide. Environmental Science and Technology 16:396–402.

Wallace, L. A. (1987) Total Exposure Assessment Methodology (TEAM) Study: Summary and Analysis, Vol. I. U.S. Environmental Protection Agency, Washington, D.C.

Water Industry Database (1992) Water Quality Profiles: Groundwater, Surface Water, Distribution System. American Water Works Association, Denver, CO: May 1992.

Weisel, C. P., and Chen, W. J. (1994) Exposure to Chlorination By-Products from Hot Water Uses. Risk Analysis 14:101–106.

Werdehoff, K. S., and Singer, P. C. (1987) Chlorine Dioxide Effects on THMFP, TOXFP, and the Formation of Inorganic By-products. Journal of the American Water Works Association Vol.79 No.9:107–113.

WHO (1993) Guidelines for Drinking-Water Quality, 2nd ed., Vol. 1, Recommendations. World Health Organization, Geneva.

3 Exposure Characteristics

Ted Johnson, P. J. (Bert) Hakkinen,
and David A. Reckhow

CONTENTS

1-57881-001-9/99/$0.00+$.50
© 1999 by International Life Sciences Institute

31

3.1   DEMOGRAPHICS

Ted Johnson

3.1.1   RELATIONSHIP TO ACTIVITY PATTERNS AND TO BUILDING AND WATER SOURCE CHARACTERISTICS

Demographics is the statistical study of human populations with respect to size, density, distribution, and vital statistics. In an exposure assessment, the analyst will typically specify the particular population of interest in terms of such demographic characteristics as residential location, housing characteristics, age, gender, work status, proximity to sources, and lifestyle. Most exposure assessments assume that demographic data can be used in some way to predict the exposure patterns of various population subgroups. In exposure assessments relating to water sources, the exposure pattern of an individual is determined by an activity pattern that typically provides a sequence of exposure events.

Each exposure event describes conditions affecting a person's exposure for a specified time interval. The exposure event may assign an individual to

- a geographic location (e.g., home census tract),
- a housing type (e.g., single family home),
- a water source (e.g., private well),
- a microenvironment (e.g., indoors — kitchen), and
- an activity (e.g., washing dishes).

Note that each of these conditions may be influenced by the demographic characteristics of the population subgroup. In particular, the demographics of the subpopulation are likely to affect the characteristics of the building where the exposure occurs and the water source. This section provides an introduction to the demographic data and its use in exposure assessments.

3.1.2 SELECTING A BASIC DEMOGRAPHIC GROUP

In performing an exposure assessment relating to water sources, it is often advantageous to organize the exposed population at three levels: the population group, the household, and the individual.

The *population group* specifies the general demographic characteristics of the population subgroup with respect to water use and exposure potential (residential location, water source, etc.). An example would be all persons living in residences with private wells that draw water from a specific aquifer.

The *household* is treated as a unit because the activities of each member of a household may affect the exposures of other members. For example, showering by one member may release water droplets into the common air breathed by all household members. In general, the residence should be treated as a physical system of time-dependent sources and sinks. The state of the system changes over time as household members move through the system and interact with it. Examples of residential parameters that affect exposure include water source, plumbing fixtures (showers, bathtubs, etc.), annual water usage, and appliances (e.g., dishwashers).

Each member of the household is treated as an *individual* with his or her own activity pattern. The sequence of exposure events in this activity pattern determines the individual's exposure over time and, indirectly, the exposures of other members of the household. The activity pattern may be affected by demographic characteristics such as age, gender, work status, food consumption, and bathing habits.

3.1.3 APPLICABLE CENSUS DATA AND SURVEY RESULTS

3.1.3.1 Bureau of Census Data

The Bureau of Census (BOC) is an important source of demographic data. The BOC conducts a national census every ten years that employs two forms: the short form and the long form. Data collected by these forms are tabulated by census units of increasing size. Blocks are typically the smallest units available for urban areas. A block is defined as an area surrounded on all sides by roads that is not intersected by a road. Blocks are aggregated into block groups, which in turn are grouped into census tracts. Rural areas are usually divided into enumeration districts, which are roughly equivalent to urban census tracts.

The long form contains questions concerning both the individuals and the household that provide a wealth of demographic data. The BOC has tabulated the data by block, block group, census tract, and enumeration district. Useful data items include the age, gender, marital status, ethnic group, education, and occupation of each household member. In addition, the long form provides the following useful data items for the household.

Number of individuals in household
Building type of residence (apartment, single family, etc.)
Value of property or rent
Years in current residence
Number of bedrooms
Complete plumbing facilities (yes/no)
Complete kitchen facilities (yes/no)
Heating fuel
Water source
Connection to public sewer
Year building constructed
Annual income of household
Utility costs

Although the long form is completed by a subset of all households, the BOC extrapolates the results to the general population in tabulating census results.

The BOC also conducts annual housing surveys that provide similar information to the data items listed above from the long form. For example, the results of the 1987 Annual Housing Survey indicate that 84.7% of U.S. housing units receive water from a public system or private company, as opposed to a well. This statistic has also been compiled for the different regions in the U.S.: 82.5% of Midwestern homes receive water from a public or private company; 93.2% in the Western region; and roughly 85% in both the Northeast and Southern regions (U.S. Bureau of Census, 1992).

Information regarding population mobility is compiled and published by the BOC. Population mobility data are important in exposure assessments because the data permit researchers to estimate the length of time a household is exposed in a particular location. A family, for example, is exposed to contaminants from a polluted private well only as long as the family resides in the local area that draws upon that polluted aquifer. If the family moves to an area supplied by a public water system, the contaminants of concern in their drinking water will be different than in their previous home (e.g., chlorine may become more of a concern). For these reasons, mobility data are essential in attempting to accurately predict a population's exposure to contaminants in drinking water.

Table 3.1.1 presents data collected by the BOC in its 1991 to 1992 Geographical Mobility study (U.S. Bureau of Census, 1993). These data refer to the number of years that a household has resided at its current location. The BOC data indicate that the majority of the population has resided in the current residence for less than 10 years, and nearly half of this majority has resided in the current home for less than 5 years. Based on the data used to create this summary table and the assumption that values are distributed evenly within each range, the authors of the *Exposure Factors Handbook* (USEPA, 1997) estimated the 50th percentile value for years living in current home to be approximately 9 years. The 90th percentile was similarly estimated to be 33 years. Accordingly, the *Handbook* proposes that a range of 9 to 33 years be used to represent the central to upper end tendency of length of residence time.

TABLE 3.1.1
Percent of Householders Living in Houses
for Specified Ranges of Time

Years Lived in Current Home	Percent of Total Householders
0–4	26.34
5–9	29.04
10–14	11.39
15–19	10.06
20–24	6.69
25–34	8.52
35–44	5.1
45–54	1.9
> 55	0.95

Source: Adapted from U.S. Bureau of the Census, 1993.

TABLE 3.1.2
Annual Geographical Mobility Rates by Type of Movement for
Selected 1-Year Periods: 1960–1992

Mobility Period	Percentage of Population Moving within the U.S.			
	Total	Same County	Same State, Different County	Different State
1960–61	20.0	13.7	3.1	3.2
1970–71	17.9	11.4	3.1	3.4
1980–81	16.6	10.4	3.4	2.8
1984–85	19.6	13.1	3.5	3.0
1989–90	17.3	10.6	3.3	3.3
1991–92	16.8	10.7	3.2	2.9

Source: Adapted from U.S. Bureau of Census, 1993.

It should be noted that the data in Table 3.1.1 refer to the *elapsed time* that a householder has lived in his or her current residence. The data do not indicate the *total time* that a person will remain in the current residence before the person dies or moves to a new residence. Estimates for this latter quantity, referred to as "residential occupancy period" (ROP), have been developed by Johnson and Capel (1992) and are discussed in Subsection 3.1.3.2.

Table 3.1.2 is derived from the same Geographical Mobility study by the BOC. This table presents annual geographical mobility rates by type of movement for selected one-year periods. As shown by the table, mobility rates in the U.S. have fluctuated but overall have tended to decrease over the last 30 years.

TABLE 3.1.3
Mobility of the Resident Population by Region: 1980

	Percent Distribution — Residence in 1975[1]			
Region	Same House	Same County (Different House)	Same State (Different House)	Different State
Northeast	61.7	22.3	8.0	6.1
Midwest	55.4	26.4	10.2	7.0
South	52.4	24.1	10.0	12.0
West	43.8	28.3	11.0	13.4
United States	53.6	25.1	9.8	9.7

[1] Survey assessed changes in residence between 1975 and 1980, including persons residing abroad in 1975.

Source: Adapted from U.S. Bureau of the Census, Statistical Abstract, 1984.

Another BOC study generated similar data for every state in the U.S., based on mobility rates between 1975 and 1980 (U.S. Bureau of Census, 1984). Table 3.1.3, adapted from this study, presents mobility as the percentages of the population in each of the four U.S. regions that moved or did not move between 1975 and 1980. These data reveal that people living in the West move most frequently, followed by the South, Midwest, and Northeast, respectively.

3.1.3.2   Other Data and Results

As discussed above, the BOC conducts the national census every 10 years. In areas with rapidly changing populations, data from the most recent census may not represent the current demographic composition of an area. To provide more up-to-date census data, some states conduct an interdecennial census of population and housing that complements the data from the decennial national census. In a recent article, Rogan (1997) emphasized the importance of such data in exposure assessment research.

The U.S. Department of Energy (U.S. DOE) also collects demographic-related data useful in exposure assessments and, in particular, housing data useful in assessing exposure via the inhalation route. The U.S. DOE's Residential Energy Consumption Survey collects data on housing characteristics, such as residential volumes, in addition to collecting information on energy use. In a 1990 survey, the U.S. DOE sampled over 5000 residences representing 94 million households nationwide (U.S. DOE, 1992). Table 3.1.4 presents a summary of indoor residential volumes estimated from this survey for major categories of housing, including a percentage breakdown of single-family homes vs. apartments. These data indicate a relationship between residential volume and both housing type and ownership. Single-family detached homes have an average volume roughly two times that of multifamily residences and mobile homes; and single-family attached homes have an average volume about halfway between these two extremes. Furthermore, within each of these housing-type categories, owner-occupied homes have about 50% greater volumes than renter-occupied homes.

TABLE 3.1.4
Average Estimated Volumes of U.S. Residences

Housing Type	Owner-Occupied		Rental		All Residences	
	Volume[1] (m³)	Percent of Total	Volume (m³)	Percent of Total	Volume (m³)	Percent of Total
Single-Family Detached	534	53.2	349	8.9	508	62.1
Single-Family Attached	436	3.9	284	2.4	378	6.4
Multifamily (2–4 units)	394	2.7	224	8.0	267	10.6
Multifamily (5+ units)	274	1.9	170	13.4	183	15.3
Mobile Home	221	4.5	177	1.1	213	5.5
All Types	494	66.2	239	33.8	408	100.0

[1] Volumes calculated from floor areas assuming a ceiling height of 8 feet.

Source: U.S. DOE 1992.

The data indicate that owner-occupied residences collectively account for two thirds of all U.S. housing.

In some cases, the type of demographic data collected by the Census Bureau and other government agencies is not directly useful in exposure and risk assessments. For example, the average total residence time is generally required for exposure assessments, not the average current residence time collected by the Census. Since survey data of total residence time do not exist, two different methods have recently been developed to model or simulate the desired distribution from available data. In one study, Israeli and Nelson (1992) used a "moving behavior model" to calculate total residence time based on current residence time data. The average total residence time calculated by this model for all U.S. households was 4.7 years. This value is less than half the average current residence time — 10.6 years — calculated by the BOC, based on Census data. Table 3.1.5 presents a summary of data for "years in one residence" used in the Israeli and Nelson study.

In another study, Johnson and Capel (1992) developed a general Monte Carlo technique to simulate a distribution of residential occupancy periods (ROPs) for a given population, based on mobility, mortality, and population data. Using this technique, Johnson and Capel estimated the mean ROP for the entire U.S. population (1987) to be 11.7 years, the median ROP to be 8.1 years, and the 95th percentile to be 33 years. The cited report by Johnson and Capel provides additional breakdowns of ROP by gender and age. Table 3.1.6 presents selected statistics for ROPs for males, females, and both genders.

The *Exposure Factors Sourcebook* (AIHC, 1994) provides the results of a simplified Monte Carlo analysis based on the Johnson and Capel approach. The reader should consult the original report by Johnson and Capel for ROP statistics rather than the *Exposure Factors Sourcebook*. The ROP statistics in the *Sourcebook* were obtained from 1000 iterations of the Monte Carlo model, whereas the original analysis by Johnson and Capel employed 500,000 iterations.

TABLE 3.1.5
Years in One Residence by Population Type

	Years Residing in One Residence						
Population Type	Min	Max	25%	50%	75%	90%	95%
All Households	0	50	0.5	1.4	3.7	12.9	23.1
Renters	0	50	0.5	1.2	2.6	5.2	8.0
Owners	0	50	1.4	5.2	17.1	32	41.4
Farms	0	60	2.4	10	26.7	48.3	58.4
Urban	0	30	0.5	1.4	3.4	10.9	21.7
Rural	0	50	1.2	3.3	9.1	21.7	32.2
Northeast U.S.	0	50	1.0	2.8	7.5	22.3	34.4
Midwestern U.S.	0	50	0.6	1.61	4.3	15.0	25.7
Southern U.S.	0	50	0.4	1.2	3.0	10.8	20.7
Western U.S.	0	50	0.4	1.2	2.9	8.9	17.1

Source: Israeli and Nelson, 1992.

TABLE 3.1.6
Descriptive Statistics for Residential Occupancy Period

	Residential Occupancy Period, Years		
Statistic	Both Genders (500,000 Persons Simulated)	Males Only (244,274 Persons Simulated)	Females Only (255,726 Persons Simulated)
Mean	11.7	11.1	12.3
10th percentile	2	2	2
25th percentile	4	3	5
50th percentile	9	8	9
75th percentile	16	15	17
95th percentile	33	31	35
99.9th percentile	59	56	61
Second Largest Value	75	73	75
Largest Value	87	73	87

Source: Adapted from Johnson and Capel, 1992.

Sedman, Funk, and Fountain (1998) developed distributions for ROP, referred to as "residence duration," by surveying a random sample of tax records identifying changes in ownership in Multnomah County, Oregon. These researchers found that the distribution of residence duration could be well-characterized by an exponential distribution with a mean of 13.6 years. Although the mean value of the distribution was found to be consistent with Johnson and Capel (1992) and with other studies, the distribution was found to differ from the national distribution developed by them. Possible causes for this disagreement include differences in geographic location (a county in Oregon vs. the nation), in housing types considered (owner-occupied vs. all types), and the time period of data (1990–1994 vs. 1990). It should also be noted that Sedman et al. did not determine the time spent in a residence by an individual selected at random, but rather the time spent by the owner of the residence. Overall, these factors reduce the general usefulness of the study results. However, the study represents the only known survey study that analyzed actual data on total time spent in a residence, rather than elapsed time.

The report by Johnson and Capel is considered the best single source of statistics concerning ROP. The Monte Carlo approach developed by Johnson and Capel is the only approach reviewed that achieves all four of the following objectives: (1) it considers total time in the residence rather than elapsed time, (2) it considers all types of housing, (3) it applies to individuals rather than households, and (4) it can be updated to use the most recent census data available. The Sedman et al. study should also be consulted when data concerning owner-occupied housing are required, particularly for locations in the Pacific Northwest.

3.1.4   REPRESENTATION OF NATIONAL VS. LOCAL POPULATION GROUPS

The population of concern in a risk assessment can vary in size from a cluster of homes drawing water from a single well to the entire population of the United States. In either case, the analyst must determine which demographic characteristics are required to perform the risk assessment and identify reliable databases that can be used as sources for this information. In general, a risk assessment involving a large population (e.g., all U.S. residents using municipal water supplies) will be performed by dividing the population into a collection of smaller, localized population groups (e.g., Houston residents using municipal water supplies). Typically, the analyst will characterize each of these local population groups by combining area-specific data obatined from national databases, such as the U.S. census, with area-specific data obtained from local sources, such as the city water department. Consequently, risk assessments will typically be concerned with local population groups at some stage of the research effort.

Ideally, a risk assessment would fully characterize each individual household included in the specified population group. This goal is difficult to achieve in practice because few databases exist that provide useful information at this level of detail. In general, the analyst should avoid subdividing the population into units smaller than census blocks. As discussed in Section 3.1.3, local population group information at the block level are readily available from the Bureau of Census. This information includes demographic data on the members of households, and on the characteristics of the residents of the area. These types of data can be utilized in exposure and risk assessments for a particular type of household (e.g., couple with one child) and residence type (e.g., apartment) in a specified block of the U.S.

Furthermore, block-level data can be combined into assessments encompassing the ranges of households and residence types in a specified neighborhood, city, county, state, or region. The level of detail of a particular assessment can be as specific as the data allow; for example, the analyst may target residences occupied by a particular combination of ethnic group, age group, and gender, with the residence itself possessing selected characteristics such as municipal water supply and two full bathrooms. These census-derived data can be combined with data on other housing characteristics (e.g., house volumes) obtained from local sources (see Section 3.3.3.2 and Murray, 1997).

In risk assessments in which water use is assumed to be affected by education or income, the analyst may wish to characterize a local population according to detailed socioeconomic factors such as highest level of school completed and occupation. Often these data are available only at the zip code level and are subject to misinterpretation. For example, the education level of people residing in a particular zip code is usually characterized by the highest level of college degree attained by the largest number of people, e.g., a four-year college degree. In addition, a particular zip code may contain people from a variety of social classes. A potential remedy for this problem is the use of the "zip + 4" data now widely available. A "zip + 4" database typically organizes data in units of 12 households (Sivadas, Matthew, and Curry, 1997).

The analyst should note that census data files organized by small spatial units (e.g., census blocks) are frequently censored to prevent users from inferring information about specific households. For example, demographic data on people who live in single-family residences may be censored at the block level when a particular block contains only one or two single-family residences. If uncensored data on this scale are mandatory for a risk assessment, the analyst may be required to perform an on-site survey to obtain the desired information.

3.1.5   CHARACTERIZATION OF SENSITIVE POPULATIONS

For the purposes of this report, populations considered "sensitive" in exposure assessments regarding drinking water are those populations that ingest more drinking water than average and engage in activities leading to greater-than-average dermal and inhalational exposures. In terms of greater-than-average ingestion exposure, for example, studies have revealed that people at higher physical exertion levels (e.g., athletes, construction workers) and/or in higher temperature climates (e.g., during summer, in desert-type regions) intake more water than the average person. In addition, pregnant and lactating women have also been found to intake a greater-than-average amount of drinking water (see Subsection 3.2.4). Less research has been conducted concerning characterization of sensitive populations for dermal and inhalational exposures to drinking water. Children are known to have higher ratios of (1) skin surface to body mass and (2) breathing volume to body mass than older population groups; consequently, dermal and inhalational exposures in children may result in a proportionally greater body burden of water contaminants associated with these exposure pathways. It may also be argued that certain subpopulations, such as athletes, engage more frequently than the average person in activities resulting in dermal and inhalational exposures (e.g., showering/bathing). Unfortunately, the available literature does not contain studies that could support or rebut such arguments.

3.1.6   DATA GAPS AND RESEARCH NEEDS

As discussed in Subsection 3.1.5, the available data relating to exposures of sensitive and highly exposed populations are primarily concerned with ingestion of water.

Little research has been conducted concerning the dermal and inhalation exposures of these special populations. Future research should focus on collecting time/activity data for occupations and lifestyles that are likely to produce high dermal and inhalation exposures. These groups may include athletes, life guards, dish washers, car washers, janitors, domestics, and other occupations that involve frequent contact with water. Water use data by age are also needed.

3.2 CONSUMPTION AND TIME/ACTIVITY DATA

Ted Johnson

A person's total exposure to water-borne pollutants over a specified time period can be determined by identifying and characterizing each event during the period that is likely to produce an exposure. The characterization of each of these "exposure events" must provide sufficient information to produce an estimate of the dose received by the exposed person. When water is consumed as a beverage, dose can be determined directly from the quantity of water consumed and the concentration of the pollutant in the water. When exposure occurs through inhalation or skin contact, dose is often estimated indirectly as a function of activity type (dish washing), location (kitchen), and duration (30 minutes) — data that are jointly referred to as "time/activity" data. In general, the analyst will require both consumption data and time/activity data to perform an in-depth exposure assessment.

This section describes the primary sources of consumption and time/activity data that are currently available. Subsection 3.2.1 lists the principal activities associated with water consumption and use. Subsection 3.2.2 focuses on activities that may result in second-hand exposure. Subsection 3.2.3 discusses differences in activity patterns among population groups. Brief descriptions of useful reports and databases are provided in Subsection 3.2.4. Two ongoing studies that are likely to produce useful data are described in Subsection 3.2.5. Subsection 3.2.6 lists the principal areas requiring additional research.

3.2.1 PRINCIPAL ACTIVITIES

The 1997 *Exposure Factors Handbook* (USEPA, 1997) lists faucet use (for drinking, cooking, general cleaning, or washing hands), showering/bathing, toilet use, clothes washing, and dishwashing as the primary types of indoor water use. In addition to direct ingestion of water from drinking and cooking, a person may be exposed to a chemical in the residential water supply through both inhalation and dermal exposure during everyday water-related activities, including showering/bathing, general cleaning, washing hands, clothes washing, and dishwashing. Table 3.2.1 presents the in-house water use rates in gallons capita per day (gcd) by type of use for eleven different studies, as summarized by the 1997 *Exposure Factors Handbook*. In terms of indoor water uses in which dermal and inhalational exposures may be of concern (i.e., excluding toilet use), these studies indicate that showers and baths generally use the greatest quantity of water, followed by laundry

TABLE 3.2.1
In-house Water Use Rates (gcd) by Study and Type of Use

Study	Total, All Uses	Shower or Bath	Toilet	Laundry	Dishwashing	Other
MWD[1]	93	26	30	20	5	12
EBMUD[2]	67	20	28	9	4	6
USDHUD[3]	40	15	10	13	2	—
Cohen[4]	52	6	17	11	18	—
Ligman[4]						
Rural	46	11	18	14	3	—
Urban	43	10	18	11	4	—
Laak[4]	42	9	20	7	4	2
Bennett[4]	45	9	15	11	4	6
Milne[4]	70	21	32	7	7	3
Reid[4]	59	20	24	8	4	3
USEPA[4]	40	10	9	11	5	5
Partridge[4]	52–86	20–40	4–6	20–30	8–10	—
Mean Across Studies[5]	59	17	18	13	6	5
Median Across Studies[5]	53	15	18	11	4	5

Sources:
[1] *Metropolitan Water District of Southern California,* 1991.
[2] *East Bay Municipal Utility Water District,* 1992.
[3] *U.S. Department of Housing and Urban Development,* 1984.
[4] Cited in Nazaroff, W. W., and Nero, A.V. (eds.), 1988.
[5] The average value from each range reported in Partridge, as cited in Nazaroff and Nero (1988), was used to calculate the median across studies. The mean and median for the "Total, all Uses" column were obtained by summing across the means and medians for individual types of water use.

and dishwashing, respectively. As the listed category "other" may include activities such as washing hands and general cleaning, it may also be of concern with respect to dermal and inhalational exposures.

Although a large water-use rate may not directly correlate with higher exposure potential, this statistic is probably indirectly related to the potential for dermal and inhalational exposures. In other words, water use rates serve as guides or clues — not direct determinants — to the indoor activities that result in the highest dermal and inhalational exposures to drinking water. This hypothesis is supported by data obtained during the National Human Activity Pattern Survey (NHAPS) as presented in Robinson and Blair (1995). During this study, researchers collected 24-hour time/activity diary data from 9386 people residing in the continental U.S. from 1992 to 1994. Levels of exposure to various water sources reported by Robinson and Blair are presented in Table 3.2.2 for the general population. Excluding activities leading to ingestion exposure (e.g., drinking tap water), Robinson and Blair report that showering/bathing were associated with the greatest number of water-related (dermal and inhalational) exposures, followed by dishwashing and clothes washing, respectively.

TABLE 3.2.2
Water-Related Exposures

Activity	Percentage of Respondents Who Engaged in Activity During Sampled Day
Took bath or shower	91%
Shower	76
Bath	15
15+ minutes (bath or shower)	31
Wash dishes	84
Dishwasher used at home	23
Washing machine used at home	43
Use humidifier	24
Swimming	8
Drank tap water	72
3+ glasses	41
Drank juice mix	61
3+ glasses	28
Drank sodas	54
3+ cans	15
Use bottled water	43
Get public water	80
Get well water	16

Source: Robinson and Blair, 1995, p.41.

3.2.1.1 Drinking Water Consumption

Contamination of drinking water originates in different ways depending on the source of the drinking water, the type of contamination, and the method of water supply delivery (i.e., type of plumbing). Groundwater, which may be used as a source of drinking water, often becomes contaminated from percolation of toxics through contaminated soil. Alternatively, contaminated runoff and direct discharges contaminate surface water, which may also be used as a drinking water source. Drinking water may become contaminated through the leaching of lead from plumbing systems. Finally, the intentional addition of substances to treat the water supply, such as chlorine, also represent a significant source of drinking water contamination in public water supply systems.

The potential dose of a toxic compound resulting from drinking water consumption is a function of consumption rate and contaminant concentration in the water. It is preferable that consumption rate be determined for the population of interest. If such population-specific data are unavailable, however, generic rates derived from relevant regional studies or national consumption surveys may be used.

Consumption surveys often report averages and distributions for both "total fluid" and "total tap water" intake. Total fluid intake is defined as "consumption of all types of fluids including tap water, milk, soft drinks, alcoholic beverages, and water intrinsic to purchased foods." Total tap water is defined as "food and beverages that are prepared or reconstituted with tap water (i.e., coffee, tea, frozen juices, soups)" in addition to straight tap water. According to the 1997 *Exposure Factors Handbook* (USEPA, 1997), intake rates based on total tap water are more representative of source-specific tap water intake than rates based on total fluid intake, for the purposes of exposure assessments. As purchased foods and beverages are widely distributed and less likely to contain source-specific water, the use of total fluid intake rates may overestimate the potential exposure to toxic substances present only in local water supplies.

Until very recently, the U.S. EPA has recommended using default drinking water intake rates of 2 liters per day for adults (USEPA, 1980). This value is a total tap water rate, as it includes drinking water consumed in the form of juices and other beverages containing tap water, such as tea and coffee. Numerous studies, however, have generated data on drinking water intake rates that support using a significantly lower default value to represent average adult drinking water consumption, while using 2 liters per day to represent the upper 80th to 90th percentile rate. Consequently, the 1997 *Exposure Factors Handbook* (USEPA, 1997) recommends 1.4 liters per day as the default drinking water value for adult consumption. The relevant studies and associated data are summarized in Subsection 3.2.4.1.

3.2.1.2 Showering

Inhalation and dermal absorption are the exposure routes of concern when assessing the exposure to contaminated water during showering (and bathing). Dermal absorption may be a significant source of exposure to contaminants in water, even when compared to ingestion. Researchers have estimated that dermal absorption of lipophilic volatile organic contaminants in drinking water accounts for an average of 64% of the dose incurred (Brown, Bishop, and Rowan, 1984). The factors influencing dermal absorption include the contaminant concentration in water, that fraction of the contaminant absorbed from the water, the duration of exposure, and surface area of the body parts exposed (USEPA, 1990).

Inhalation of volatilized contaminants in domestic water during showering is another important route of exposure. Factors that influence inhalation exposure include the volatility of a contaminant, the duration of exposure, and the inhalation rate of the exposed person(s). Researchers have simulated the daily exposure of people to volatilized organic chemicals in domestic water, and have observed that showering delivers inhalation dose at the greatest rate, compared to other indoor water uses. Furthermore, inhalation dose was found to be significant when compared to doses resulting from ingestion (Wilkes, Small, Davidson, and Andelman, 1996).

3.2.1.3 Laundry

In households that use washers and dryers, laundry-related exposures tend to occur indirectly through inhalation of contaminants in the water that volatilizes as clothes are washed and dried. Direct dermal exposures and inhalation exposures also occur when clothes are washed by hand, and moved from a washing machine to a dryer or to a clothes line. Hakkinen (1993) provides an overview of the types of exposures that can occur through use of laundry products and provides guidelines for estimating exposures resulting from skin contact, inhalation, and ingestion. In general, these exposures have not been well-characterized in the scientific literature.

3.2.1.4 Dish Washing

Direct dermal exposures occur as dishes are washed, rinsed, and dried by hand. Inhalation exposures can also occur as contaminants in the water in the sink or dishwasher are volatilized. The Hakkinen (1993) reference cited above provides guidance for estimating exposures associated with dish washing.

3.2.1.5 Other Activities

Any cleaning or bathing activity that uses water may produce dermal and inhalation exposures. Typical household cleaning activities include washing floors, windows, cars, pets, and outdoor surfaces. In addition to household bathing in tubs and showers, bathing-type activities occur in swimming pools, hot tubs, jacuzzis, and saunas. The Hakkinen (1993) reference cited previously provides guidance for estimating exposures associated with hand washing, hard surface cleaning, and other cleaning activities.

3.2.2 INDIRECT (SECOND-HAND) EXPOSURES

The water-related activities of each household member may affect the exposures of other members. Activities such as showering, clothes washing, and dish washing may release volatile compounds and water droplets into the residential air, where they can be respired by all members of the household. As most of the existing time/activity databases track the location of only one household member at a time, it is difficult to characterize the indirect exposures of other members. Future research in this area should attempt to collect simultaneous time/activity data for all household members.

3.2.3 VARIATION AMONG POPULATION GROUPS

Domestic water use varies among population groups. Since water usage studies typically focus on water consumption rates, most of the data available on variations across population groups also tend to focus on differences in water consumption. For example, the 1997 *Exposure Factors Handbook* (USEPA, 1997) presents the results of several studies showing water intake rates are less for adult women on average than for adult men. In addition, daily consumption of water varies with levels of physical activity and fluctuations in temperature and humidity (National Academy of Science, 1977). It is therefore reasonable to assume that individuals in physically demanding occupations or in warmer regions intake more water on average than other individuals. To obtain accurate exposure assessments, analysts should assign such population groups a higher water intake rate than the 1.4 L/day value proposed for adults in general. McNall and Schlegel (1968) provide relevant data obtained from exercise and temperature studies that positively link higher water intake rates with higher temperature and exertion levels. According to these researchers, hourly intake rates can range from 0.21 to 0.65 liters per hour depending on the temperature and activity level. Research by the U.S. Army (1983) indicates that intake among physically active individuals can range from 6 L/day in temperate climates to 11 L/day in hot climates.

Ershow, Brown, and Cantor (1991) provide estimates for total fluid intake and total tap water intake among pregnant and lactating women (ages 15 to 49 years). Water intake rates for 188 pregnant women, 77 lactating women, and 6201 nonpregnant, nonlactating control women were evaluated. It was observed that lactating women had the highest mean total fluid intake rate (2.24 L/day) compared with both pregnant women (2.08 L/day) and control women (1.94 L/day). Lactating women also exhibited a higher mean total tap water intake rate (1.31 L/day) than pregnant women (1.19 L/day) and control women (1.16 L/day). These same researchers also reported that a sample of 1885 rural women consumed more total water (1.99 L/day) and tap water (1.24 L/day) than a sample of 4581 urban/suburban women, who consumed on average 1.93 L/day total water and 1.13 L/day of tap water.

The 1997 *Exposure Factors Handbook* (USEPA, 1997) notes that water intake differs significantly by region. Total water and tap water intake rates are found to be lowest in the northeastern region of the United States (1.82 and 1.03 L/day) and highest in the western region of the United States (2.06 L/day and 1.21 L/day).

Reports by Robinson and Blair (1995) and by Tsang and Klepeis (1996) provide data from the National Human Activity Pattern Survey (NHAPS) study, which can be used to characterize the variation of water-related activities by population group. Table 16 of the Robinson and Blair report lists the percentages of persons in specific population groups that engaged in selected activities. The population groups are categorized by a number of useful demographic factors, including gender, age, race, education, day of week, calendar quarter, region, and housing type. Robinson and Blair noted the following patterns in the tabulated results:

> *Race* (White, African-American, Asian, Hispanic, other). African-Americans were slightly less likely to drink juice mixed with tap water, to drink well water, and to have dishwashers and washing machines operating while at home. They were more likely to report taking a bath and drinking bottled water. Asian-Americans were least likely to drink carbonated beverages and more likely to report taking showers and drinking both bottled water and publicly supplied water. Hispanics were higher in soda consumption, but lower in consuming juices mixed with water and using dishwashers and humidifiers.
>
> *Education* (less than 12 years, high school, some college, college, postgraduate). The high-school-educated were more likely to report baths, compared to the college educated, who take showers. The less educated were more likely to drink tap water and water in fruit juices; they were also more likely to use well water. College-educated people were more likely to drink bottled water.

Gender. Men were slightly more likely to take showers, and women were more likely to take baths. Women were more likely to be at home when a washing machine was being used or to have washed dishes.

Age. Preschool children were more likely to take a bath than a shower, and were more likely to be home when dishwashers and portable humidifiers were being used. Teens reported the highest levels of pool swimming, soda consumption, use of well water, and home washing machine exposure. Young adults were high in terms of taking showers, drinking bottled water, and having access to public water. Older adults were more likely to take baths than showers; they also reported higher consumption of home tap water and juices mixed with tap water.

Day of the week. Sundays were associated with higher levels of taking showers, washing machine use, and dishwasher use.

Housing type (apartment, house, townhouse, other). Residents of apartments and townhouses were more likely to drink bottled water and to have access to public water than were residents of houses.

Tsang and Klepeis (1996) have analyzed data from the NHAPS study and prepared an extensive set of tables listing frequency and duration statistics for a wide variety of data classifications. In each case, the statistics are tabulated by gender, age, race, employment status, day of week, season, and disease group (asthma, angina, and bronchitis/emphysema).

3.2.4   APPLICABLE DATABASES

Table 3.2.3 provides brief summaries of 32 reports and databases that present useful water consumption and time/activity data. Subsections 3.2.4.1 to 3.2.4.4 provide a more detailed discussion of the characteristics and potential uses of these resources.

3.2.4.1   Water Consumption

More studies have been conducted concerning water consumption than of any other domestic water use. Ten different water consumption surveys described in the 1997 *Exposure Factors Handbook* (USEPA, 1997) are summarized in this section. Although these studies were performed independently and were based on different populations, the resulting mean and upper-percentile intake rate estimates appear to be relatively consistent across the studies. For example, the estimates for the average drinking water intake rate of adults obtained from these studies generally falls between 1.30 and 1.40 L/day. Based on these studies, the 1997 *Exposure Factors Handbook* recommends that a value of 1.4 L/day be used for exposure assessments. Furthermore, the (previously) widely used default adult drinking water consumption rate of 2 L/day should now be used to represent the upper 80th to 90th percentile of intake rates among the adult population. For children less than 1 year old, the *Handbook* recommends an intake rate of 0.3 L/day for water-based beverages and 0.7 L/day for the upper-percentile rate. An average drinking water intake rate of 0.7 L/day is recommended for children aged 1 to 10 years, with an upper-percentile rate

TABLE 3.2.3
Brief Summaries of Reports and Databases Providing Useful Water Consumption and Time/Activity Data

Citation	Title	Type of Research	Scope	Capsule Summary	Comments
Cantor et al., 1987	Bladder cancer, drinking water source, and tap water consumption: a case control study.	Questionnaire survey	8000 white adults, 21–84 years of age	Subjects were asked to recall beverage intake over typical week.	Winter only. Not representative of general U.S. population. Recall method unreliable.
Canadian Ministry of National Health and Welfare, 1981	Tap water consumption in Canada.	Questionnaire survey	1000 participants	Monitored water intake over two-day period. Tap water intake rates as a function of physical activity.	Representative survey of Canadian population. Data tabulated by age group and season.
Ershow and Cantor, 1989	Total water and tap water intake in the United States: population-based estimates of quantities and sources.	Intake rate estimation		Estimates of water intake rates based on data collected by the USDA Nationwide Food Consumption Survey.	Representative sample of U.S. population. Seasonally balanced. Based on recall of use.
Roseberry and Burmaster, 1992	Lognormal distribution of water intake by children and adults.	Intake rate estimation		Lognormal distributions were fit to Ershow and Cantor's water intake data and population-wide distributions for total fluid and total tapwater intake rates were estimated.	See Ershow and Cantor, 1989.
National Academy of Science, 1977	Drinking water and health, volume 1.	Intake rate estimation		Results of literature survey were used to estimate average per capita liquid consumption.	Results of limited use because term "liquid" was not clearly defined.
Pennington, 1983	Revision of the total diet study food list and diets.	Estimation of average intake rates		Estimates of average intake rates for foods and beverages based on FDA's Total Diet Study and NHANES II.	See Ershow and Cantor, 1989.

Reference	Title	Type	Sample	Details	Comments
Gillies and Paulin, 1983	Variability of mineral intakes from drinking water: a possible explanation for the controvery over the relationship of water quality to cardiovascular disease.	Survey	109 adults in New Zealand	Subjects collected duplicate samples of consumed water.	Short-term survey. New Zealand population may not represent U.S.
Hopkins and Ellis, 1980	Drinking water consumption in Great Britain: a survey of drinking habits with special reference to tap-water-based beverages.	Questionnaire and diary survey	3546 participants in England, Scotland, and Wales	Type and quantity of beverages consumed over one week.	Short-term survey. British population may not represent U.S.
U.S. EPA, 1984	An estimation of the daily average food intake by age and sex for use in assessing the radionuclide intake of individuals in the general population.			Total tap water intakes were estimated by combining the average daily intakes of tap water, water-based drinks, and soups from USDA data.	See Ershow and Cantor, 1989.
ICRP, 1981	Report of the task group on reference man.	Literature summary		Summarized data on fluid intake levels.	Results are consistent with other studies.
U.S. EPA, 1992	Dermal exposure assessment: principles and applications.	Exposure factors manual		Provides default frequencies for showering/bathing.	
U.S. EPA, 1989	The exposure factors handbook	Exposure factors manual		Data on contact time and frequency of bathing.	
James and Knuiman, 1987	An application of Bayes methodology to the analysis of diary records from a water use study.	Diary survey	2500 households in Australia	Diary records used to estimate showering and bathing habits of Australian population	Australian population may not be representative of U.S.
Tarshis, 1981	The "Average American" Book.	Summary of data from small surveys and other sources.		Provides statistics on bathing frequencies for Americans.	Data may be out of date.

TABLE 3.2.3 (continued)
Brief Summaries of Reports and Databases Providing Useful Water Consumption and Time/Activity Data

Citation	Title	Type of Research	Scope	Capsule Summary	Comments
U.S. EPA, 1997	The Exposure Factors Handbook.	Exposure factors manual.		In-house water usage for shower or bath based on 11 studies.	
Metropolitan Water District of Southern California, 1991	Urban water use characteristics in the metropolitan water district of southern California.	Compilation of water use data.		Substantial information on water use in California households.	California data may not represent U.S.
U.S. Dept. of Housing and Urban Development, 1984	Residential water conservation projects: summary report.		200 households	Water use was monitored for a 20-month period in each household	
Nazaroff and Nero, 1988	Radon and its decay products in indoor air.	Compilation of data from several small surveys.		Data on reported water usage for shower/bath, toilet, laundry, dishwashing, and other (e.g., faucets).	
U.S. EPA, 1997	The Exposure Factors Handbook.	Exposure factors manual.		In-house water usage for laundry based on 11 studies.	
U.S. EPA, 1988	Superfund Exposure Assessment Manual.	Exposure factors manual.		Data quantifying contact time and frequency of swimming based on information from Bureau of Outdoor Recreation.	
U.S. EPA, 1992	Dermal exposure assessment: principles and applications.	Exposure factors manual.		Provides default frequencies for swimming.	
Robinson, 1977	Changes in Americans' Use of Time: 1965–1975 — a Progress Report.	Two diary surveys.	Respondents: 1244 (1965) and 1500 (1975).	Data on dishwashing, laundry, and bathing.	Dated information. Statistical analysis of data not provided.
Juster et al., 1983	Study Description: 1975–1981 Time Use Longitudinal Panel Study.	Three diary surveys.	National surveys in 1965–66, 1975–76, and 1981.	Adds 1981 study to two diary surveys described by Robinson, 1977.	Data not directly applicable to exposure assessments.

Reference	Title	Study Type	Sample	Description	Limitations
Timmer, Eccles, and O'Brien, 1985	How children use time.	Analysis of diary data.	Diary and interview were administered to children.	Study of children's activity patterns is follow-up to 1975–76 surveys (Robinson, 1977).	Dated information. Omits summer activities.
Hill, 1985	Patterns of time use.	Analysis of diary data.		Analysis of 1975–76 diary data collected by Robinson.	Seasonally balanced data. Omits children. Dated information.
Robinson and Thomas, 1991	Time Spent in Activities, Locations, and Microenvironments: a California-National Comparison Project Report.	Analysis of diary data.		Compared 1987–88 CARB study with 1985 national study. Data on dishwashing, laundry, and bathing. Dated information.	CARB data represents California only. Recall diaries were used in both studies.
CARB, 1991	Study of children's activity patterns.	Diary study.	1200 children in California (0–11 years)	24-hour recall diary provided detailed data on activities and locations.	Recall diary. Provides data by age, gender, and season. May not represent U.S.
Tarshis, 1981	The "Average American" Book.	Summary of data from small surveys and other sources.		Data on time spent on personal grooming.	Data may be out of date.
U.S. EPA, 1992	Dermal Exposure Assessment: Principles and Applications.			Data on showering/bathing times.	
James and Knuiman, 1987	An application of Bayes methodology to the analysis of diary records from a water use study.			Data on shower durations.	Australian population may not be representative of U.S.
Robinson and Blair, 1995	Estimating Exposure to Pollutants Through Human Activity Pattern Data.	Diary survey.	9386 people in U.S.	24-hour recall diary used to obtain detailed data on water consumption, use of various appliances, cleaning activities, and bathing.	Large national sample. Study focused on exposure-related activities.

TABLE 3.2.3 (continued)
Brief Summaries of Reports and Databases Providing Useful Water Consumption and Time/Activity Data

Citation	Title	Type of Research	Scope	Capsule Summary	Comments
Tsang and Klepeis, 1996	Descriptive Statistics Tables from a Detailed Analysis of the National Human Activity Pattern Survey (NHAPS) Data	Statistics derived from the study by Robinson amd Blair.	9386 people in U.S.	Comprehensive tabulation of time/activity statistics derived from the study by Robinson and Blair	Statistics are tabulated by demographic groups and other useful classification factors.
Johnson et al., 1992	A comparison of ten time/activity databases: effects of geographic location.	Comparison of ten time/activity databases.		Provides descriptive statistics for time spent in various microenvironments.	No data presented on water use.
DeOreo, Heaney, and Mayer, 1996	Flow trace analysis to assess water use.	Measured flow rates	16 homes in Boulder, CO.	Data on nine categories of use (toilet flushings, bathing, clothes washing, irrigation, etc.).	Highly accurate data for a small sample of homes. Indicates that flows may vary widely within a neighborhood.

of 1.3 L/day. For children aged 11 to 19 years, the *Handbook* recommends an average intake rate of 1.0 L/day, with an upper-percentile value of 1.7 L/day (USEPA, 1997). The 1994 *Exposure Factors Sourcebook* (AIHC, 1994) recommends similar water intake rates based on its survey of the studies listed in the 1989 *Exposure Factors Handbook* (USEPA, 1989).

The 1997 *Exposure Factors Handbook* notes that all currently available drinking water intake studies are based on short-term survey data. Such short-term data are generally suitable for estimating mean intake values that are representative of both short- and long-term consumption patterns. Short-term data, however, are not suitable for representing upper-percentile long- term consumption patterns because more variability generally occurs in short-term surveys. Another significant limitation to the drinking water surveys currently available is that most are based on subject recall rather than the use of diaries. This limitation introduces a source of uncertainty into the estimated intake rates because of the subjective and fallible nature of a recall-based survey.

The 1997 *Exposure Factors Handbook* separates the currently available drinking water surveys into two categories: "key general population studies" and "other relevant general population studies." Following are brief summaries of the studies designated as "key" by the *Handbook*:

- Cantor, K. P., Hoover, R., Hartge, P., Mason, T. J., Silverman, D. T. et al., Bladder cancer, drinking water source, and tap water consumption: a case-control study, *J. Natl. Cancer Inst.*, 79(6), 1269–1279, 1987. The National Cancer Institute (NCI) conducted this drinking water study as part of an investigation into the possible relationship between bladder cancer and drinking water. The survey was a population-based, case control study during which approximately 8000 adult white individuals, 21 to 84 years of age (2805 cases and 5258 controls) were interviewed using a standardized questionnaire. The subjects attempted to recall their "level of intake" of tap water and other beverages in a typical week during the winter prior to the interview. Average intake rates for a typical week were compiled by sex, age group, and geographic region, based on responses from 5258 white controls (3892 males; 1366 females). The average tap water intake rate was 1.4 L/day for men, which represented 70% of the total fluid intake rate (2.01 L/day). The average tap water intake rate for women was 1.35 L/day, which represented 79% of the total fluid intake (1.72 L/day). The overall average adult tap water intake rate estimated from this study was 1.39 L/day. Limitations to this study include: the population surveyed was not representative of the general U.S. population; the data were obtained through recall; the data represent a winter period when intake rates may be relatively low; and the short-term nature of the survey makes extrapolation to long-term consumption patterns difficult.
- Canadian Ministry of National Health and Welfare, Tapwater Consumption in Canada, Document No. 82-EHD-80, Public Affairs Directorate, Department of National Health and Welfare, Ottawa, Canada, 1981. The Canadian Department of Health and Welfare conducted this study in an

effort to determine the per capita total tap water intake rates for various
age/sex groups during winter and summer seasons. Approximately 1000
individuals were surveyed, and intake rates were evaluated as a function
of physical activity. In addition, researchers attempted to obtain a representative sample of the Canadian population. Subjects were instructed to
monitor their water intake for a 2-day period (1 weekday, and 1 weekend
day) in both the summer and winter during 1977 and 1978. Survey results
for adults over 18 years old indicate that the average total tap water intake
rate was 1.49 L/day and that the 90th percentile rate was 2.50 L/day. The
daily total tap water intake rates for all ages and seasons combined was
estimated to average 1.34 L/day, with a 90th percentile rate of 2.36 L/day.
This study is useful in that it provides intake rate data based on different
age groups and seasonal variability. The survey also included data for
some tap-water-containing items not covered by other studies (including
ice cubes, popsicles, and infant formula) and may therefore be more
representative of total tap water consumption than other less comprehensive surveys. Limitations of the study include the use of subjectively
determined volume (container) sizes of fluid; the underrepresentation of
certain age groups; the short-term nature of the survey; and the possibility
that Canadian water consumption rates may not be indicative of U.S. water
consumption rates.

- Ershow, A. G., and Cantor, K. P., Total Water and Tapwater Intake in the
  United States: Population-Based Estimates of Quantities and Sources, Life
  Sciences Research Office, Federation of American Societies for Experimental Biology, 1989. This study estimated water intake rates based on
  data collected by the USDA during the 1977–1978 Nationwide Food
  Consumption Survey (NFCS). Tapwater and total water daily intake rates
  were calculated by age for males, females, and both sexes combined. Total
  tap water intake ranged from approximately 0.15 to 3.78 L/day with a
  mean intake rate of 1.37 L/day for adults (ages 20 to 64). For children
  (ages 1 to 10), total tap water intake ranged from approximately 0.06 to
  1.95 with a mean intake rate of 0.74 L/day. The 90th percentile rate for
  adults was 2.27 L/day, and for children 1.29 L/day. One advantage of this
  survey is that it was based on the USDA NFCS, which is a large geo-
  graphically and seasonally balanced survey of a representative sample of
  the U.S. population. A limitation to the study however, is the collection
  of data through short-term recall.

- Roseberry, A. M., and Burmaster, D. E., Lognormal distributions for water
  intake by children and adults, *Risk Analysis*, 12, 99–104, 1992. In this study,
  researchers fit lognormal distributions to water intake data obtained from
  Ershow and Cantor (1989). In addition, researchers estimated population-
  wide distributions for total fluid and total tap water intake rates based on
  proportions of the population in each age group. The mean total tap water
  intake rates were estimated to be 1.27 L/day for adults aged 20 to 65 years,
  and 1.34 L/day for adults 65 years or older. As these intake rates were based

on the data originally presented by Ershow and Cantor (1989), the same advantages and disadvantages associated with that data apply to this study.

The following are brief summaries of the water consumption surveys designated as "Other Relevant General Population Studies" by the U.S. EPA in the 1997 *Exposure Factors Handbook* (USEPA, 1997):

- National Academy of Sciences (NAS), *Drinking Water and Health, Vol. 1,* National Academy of Sciences–National Research Council, Washington, D.C., 1977. NAS surveyed nine different literature sources and calculated the average per capita water (liquid) consumption to be 1.63 L/day. NAS nonetheless adopted a larger volume (2 L/day) to represent the intake of "the majority of water consumers." The results of this study are of limited use in recommending total tap water intake rates because the use of the term "liquid" was not clearly defined.
- Pennington, J. A. T., Revision of the total diet study food list and diets, *J. Am. Diet. Assoc.,* 82, 166–173, 1983. This study reports average intake rates for various foods and beverages for five age groups of the population, based on data from the Total Diet Study conducted by the U.S. Food and Drug Administration. Representative diets were developed based on 24-hour recall and 2-day diary data obtained from the 1977–1978 Nationwide Food Consumption Survey (NFCS) and 24-hour recall data from the Second National Health and Nutrition Examination Survey (NHANES II). Total fluid and total tap water intake rates for five age group were calculated based on the consumption rates for certain food categories. The average adult intake rate for total tap water was estimated to be 1.15 L/day, based on the average of the intake rates for adults aged 25 to 30 (1.04 L/day) and 60 to 65 (1.26 L/day). Because these intake rates are based on the USDA NFCS, the same limitations associated with the Ershow and Cantor (1989) data apply to these data.
- Gillies, M. E., and Paulin, H. V., Variability of mineral intakes from drinking water: a possible explanation for the controversy over the relationship of water quality to cardiovascular disease, *Int. J. Epid.,* 12(1), 45–50, 1983. This study was conducted to evaluate variability of mineral intake from drinking water in a New Zealand study population of 109 adults (75 females, 34 males), ranging in age from 16 to 80 years (mean age of 44 years). The study subjects collected duplicate samples of any water they consumed directly from the tap or used in beverage preparation during a 24-hour period. The mean total tap water intake rate was 1.25 (±0.39) L/day for this study population, the median total tap water intake rate was 1.26 L/day, and the 90th percentile rate was 1.90 L/day. The recorded intake rates ranged from 0.26 to 2.80 L/day. Limitations to this study include the fact that the data were based on a short-term survey, and the possibility that the New Zealand population's consumption habits are not representative of U.S. consumption habits.

- Hopkins, S. M., and Ellis, J. C., Drinking Water Consumption in Great Britain: A Survey of Drinking Habits with Special Reference to Tap-Water-Based Beverages, Technical Report No. 137, Water Research Centre, Wiltshire, Great Britain, 1980. This study estimated the drinking water consumption rates of 3564 individuals from 1320 households in England, Scotland, and Wales, and was conducted in Great Britain over a 6-week period (during September and October, 1978). Participant selection was random. Subjects completed a questionnaire and diary indicating the type and quantity of beverages consumed over a 1-week period. The mean per capita value for total tap water intake rate based on all individuals surveyed was 0.95 L/day, while the total liquid intake rate was 1.59 L/day. Intake rates were also estimated for males and females in various age groups. The use of diary recording in this study is an advantage over the more typical studies, which rely on recall methods. Limitations include the fact that the study was based on a short-term survey and the possibility that Great Britain's drinking patterns are not representative of the U.S.'s drinking patterns.
- U.S. Environmental Protection Agency, An Estimation of the Daily Average Food Intake by Age and Sex for Use in Assessing the Radionuclide Intake of Individuals in the General Population, Report No. EPA-520/1-84-021, 1984. This study conducted by the U.S. EPA assessed radionuclide intake through food consumption, based on data collected by the USDA in the 1977–78 NFCS. As part of this study, the U.S. EPA determined daily food and beverage intake levels by various age groups. Mean tap water intake rates ranged from 0.62 to 0.76 L/day for adults aged 20 and older. Total tap water intake rates were estimated by combining the average daily intakes of tap water, water-based drinks, and soups for each age group. Mean total tap water intake rates ranged from 1.04 to 1.47 L/day for adults (20 and older), and mean intake rates ranged from 0.19 to 0.90 L/day for children (up to 19 years old). As these intake rates are based on the same USDA NFCS data used in Ershow and Cantor (1989), the same data limitations apply.
- International Commission on Radiological Protection (ICRP), *Report of the Task Group on Reference Man*, Pergamon Press, New York, 1981. This reference document presents summary data on fluid intake levels. The authors state that drinking water intake rates range from about 0.37 L/day to about 2.18 L/day, for tap water and water-based drinks consumed by adults under "normal" conditions. For children, the intake rate ranges from 0.54 to 0.79 L/day. Data presented in this document are generally consistent with data from other sources.

3.2.4.2 Showering/Bathing

As the time spent showering or bathing is an important factor in calculating exposure via both dermal absorption and inhalation, researchers have conducted a number of studies and surveys to determine reasonable estimates of these quantities. The available data are not nearly as extensive as the data available on water consumption estimates, however. Each of the following studies was cited by the 1997 *Exposure Factors Handbook* (USEPA, 1997) or the 1994 *Exposure Factors Sourcebook* (AIHC, 1994) as being particularly useful.

- U.S. Environmental Protection Agency, Dermal Exposure Assessment: Principles and Applications, Report No. EPA/600/891/011B, Office of Health and Environmental Assessment, Washington, D.C., 1992. This 1992 manual on dermal exposure assessment provides a range of recommended defaults for dermal exposure factors applicable to showering/bathing. The "central" defaults for event duration and frequency for bathing and showering are 10 min/event, 1 event/day, and 350 days/yr. The manual recommends a central default value for exposure duration at the same residence of 9 years. The "upper" default values for event time and frequency for bathing and showering are 15 min/event, 1 event/day, and 350 days/yr. The recommended upper default value for exposure duration at the same residence is 30 years.
- U.S. Environmental Protection Agency, the Exposure Factors Handbook, Office of Health and Environmental Assessment, Washington, D.C., 1989. This 1989 version of the *Exposure Factors Handbook* reports data quantifying the contact time and frequency of bathing. The 50th, 75th, and 90th percentiles of bathing time are approximately 7, 9, and 12 min/day, respectively. Ninety percent of Americans report bathing once a day, and 5% report bathing more than once a day.
- James, I. R., and Knuiman, M. W., An application of Bayes methodology to the analysis of diary records from a water use study, *Journal of the American Statistical Association,* 82(399), 705–711, 1987. This study, conducted in Australia, used diary records of 2500 households in an effort to determine the showering and bathing habits of the Australian population. Shower duration was found to have a median value of 6.8 minutes, with a 90th percentile duration of approximately 12 minutes. [The *Exposure Factors Sourcebook* (AIHC, 1994) recommends the median listed by James and Knuiman (6.8 minutes per shower) as the preferred reference value to be used in risk assessments.] Shower flow rates were estimated as ranging from 5 to 15 gallons (18.9 to 86.8 liters) per minute. A limitation of this study is that Australian shower durations may not be representative of U.S. shower durations.
- Tarshis, B., The "Average American" Book, New American Library, New York, 1981. This book addresses the habits, tastes, lifestyles and attitudes of the American people, including personal grooming. The personal grooming data reported in this document were gathered from small surveys, the Newspaper Advertising Bureau, and magazines. The following useful statistics are provided: 90% of all Americans take some sort of a bath in an average 24-hour period; 5% average more than 1 shower or bath a day; 75% of men shower, 25% take baths; 50% of women take showers, 50% take baths; 65% of teenage girls 16 to 19 shower daily; 55% of teenage girls take at least one bath a week; 50% of women use

an additive in their bath every time they bathe; younger and richer people are more likely to shower than bathe; and showering is more popular than bathing in large cities. Limitations of this study concern the age and source of the data; the data were compiled from small surveys, newspapers, and magazines available prior to 1981. As such, the data may not reflect current trends of the general population.

- U.S. Environmental Protection Agency, The Exposure Factors Handbook, EPA/600/P-95/002Fc, Washington, D.C., 1997. An excellent general reference, the 1997 Exposure Factors Handbook reports that the mean in-house water usage for "shower or bath," based on 11 different studies, is 17 gallons per capita per day (with a median of 15 gcd).

The studies summarized above do not provide distributions of shower duration by age and gender. This data gap is unfortunate, as the *Exposure Factors Sourcebook* indicates that age and gender significantly affect shower duration.

3.2.4.3 General (Other Studies)

- (a) Metropolitan Water District of Southern California, Urban water use characteristics in the metropolitan water district of Southern California. Draft Report, August and (b) East Bay Municipal Utility Water District, Urban Water Management Plan, written communication to J. B. Andelman, July 1992. These two studies collected substantial information on water use in California households.
- U.S. Department of Housing and Urban Development, Residential Water Conservation Projects: Summary Report, Report No. HUD-PDR-903, Office of Policy Development and Research, Washington, D.C., 1984. This study by the U.S. Department of Housing and Urban Development monitored water use in 200 households over a 20-month period.
- Nazaroff, W. W., and Nero, A.V. (eds.), *Radon and Its Decay Products in Indoor Air,* John Wiley & Sons, New York, 1988. This study assembled the results of several smaller surveys, typically involving between 5 and 50 households each. These studies reported water usage in gallons per capita per day (gcd) by type of use — shower/bath, toilet, laundry, dishwashing, and other (e.g., faucets).
- U.S. Environmental Protection Agency, *The Exposure Factors Handbook,* EPA/600/P-95/002Fc, Washington, D.C., 1997. The 1997 *Exposure Factors Handbook* reports that the mean in-house water usage for laundry, based on 11 different studies, is 13 gallons per capita per day (with a median of 11 gcd); the mean in-house water usage for "dishwashing," based on 11 different studies, is 6 gallons per capita per day (with a median of 4 gcd); and the mean in-house water usage for "toilet," based on 11 different studies, is 18 gallons per capita per day (with a median of 18 gcd).
- U.S. Environmental Protection Agency, Superfund Exposure Assessment Manual, Report No. EPA/540/1-88/001, NTIS PB89-167985, Office of Emergency and Remedial Response, Washington, D.C., 1988.

This document reports data quantifying the contact time and frequency of swimming, based on information from the Bureau of Outdoor Recreation. The average contact time for swimming is 2.6 hours/day, and the average frequency is 7 days/yr. These averages are likely to vary significantly depending on local climate and location, however.

- U.S. Environmental Protection Agency, Dermal Exposure Assessment: Principles and Applications, Report No. EPA/600/891/011B, Office of Health and Environmental Assessment, Washington, D.C., 1992. This document provides a range of recommended defaults for dermal exposure factors associated with swimming. Central default values for event time and frequency for swimming are 0.5 hr/event, 1 event/day, and 5 days/yr. The central default value for duration of exposure at the same location is 9 years. The upper-range default values for event time and frequency for swimming are 1.0 hr/event, 1 event/day, and 150 days/yr; the upper-range default value for exposure duration is given as 30 years at the same location. For swimming and bathing scenarios, the handbook notes that past exposure assessments have assumed that 75% to 100% of the skin surface is exposed.

3.2.4.4 Time/Activity Data

The long-term pollutant dose received by a person is dependent on the accumulated dose received over a series of shorter-term exposure "events." These events may include drinking fruit juice, taking a shower, or washing the car. The dose received during each event is determined by the pollutant concentration to which the person is exposed, the event duration, and the uptake rate of the pollutant into the body. As human time/activity data are useful in developing reasonable sequences of exposure events, researchers have conducted a number of studies to acquire these data. The 1997 *Exposure Factors Handbook* (USEPA, 1997) summarizes the majority of these studies, most of which are useful in exposure assessments. Those studies containing data useful in predicting domestic water use patterns are discussed below. A recently completed national time/activity study conducted by Robinson and Blair (1995) is also discussed here.

- Robinson, J. P., Changes in Americans' Use of Time: 1965–1975 — A Progress Report, Cleveland State University: Communication Research Center, Cleveland, Ohio, 1977. This survey compares time use data from a national survey conducted in 1965–1966 and one conducted in 1975. Both surveys employed time-diary methods to collect the data. There were 1244 respondents in the 1965–1966 study, and over 1500 respondents in the 1975 study. The surveys collected data on time spent with family members, time spent at various locations during activities, and time spent performing primary and secondary activities. The various activities were classified according to 96 categories, combined into five major categories, and presented by gender, age, marital and employment status, race, and education. In these studies, two of the major five categories were "personal care" and "family care," both presented in hours/week. Time estimates for these major activity categories were derived by combining various minor activity categories, including: "doing dishes, rinsing dishes, loading and unloading dishwasher"; "laundry and clothes care — wash"; and "washing, showering, bathing." Limitations to these studies include the dated nature of the survey information (which is now 20 to 30 years old), and the fact that detailed statistical descriptions of the data sets were not provided.

- Juster, F. T., Hill, M. S., Stafford, F. P., and Parsons, J. E., Study Description: 1975–1981 Time Use Longitudinal Panel Study, Survey Research Center, Institute for Social Research, University of Michigan, Ann Arbor, MI, 1983. This document describes a group of studies known as the Time Allocation data series, which began with a survey in 1965–1966 as part of a multinational project. A second national time use study was conducted in 1975–1976, and a third in 1981. The surveys included representative samples from the adult population (18 years and older) and from children between the ages of 3 and 17 in the U.S. Twenty-four hour recall diaries were completed by respondents and their spouses. The surveys gathered information regarding employment status; earnings and other income and benefits; health, friendships, and associations of the respondents; stock technology available to the household, house repair and maintenance activities of the family; division of labor in household work and related attitudes; physical characteristics of the respondents' housing structure; net worth and housing values; job characteristics; and characteristics of mass media usage on a typical day. A fundamental limitation of these surveys is that they do not present time use data that are directly relevant to exposure assessments. The information collected, however, can be used to assess various exposure pathways and scenarios associated with human activities.

- Timmer, S. G., Eccles, J., and O'Brien, K., How children use time, in *Time, Goods, and Well-Being*, Juster, F. T., and Stafford, F. P., Eds., Survey Research Center, Institute for Social Research, University of Michigan, Ann Arbor, MI, 1985, 353–380. This study of children's activity patterns is based on 1981–1982 follow-up data collected from households included in an earlier 1975–1976 study. Children between the ages of 3 and 17 were interviewed twice (first in person, then over the phone). Questionnaires consisting of a time diary and a standardized interview were administered to the children with the help of parents. In the time diary, children's activities beginning at 12:00 a.m. the previous night were recorded, including: the duration and location of each activity, the presence of another individual, and whether they were performing other activities at the same time. "Personal care" and "household work" are two of the major activity groups for which daily duration (minutes/day) data are presented. Limitations to this study include the fact that the collected data represent only the time of year when children are in school, and the likelihood that the

1981–1982 data do not represent current conditions. One major advantage of this study is that diary recordings of activity patterns were not completely based on recall.

- Hill, M. S., Patterns of time use, in *Time, Goods, and Well-Being,* Juster, F. T., and Stafford, F. P., Eds., Survey Research Center, Institute for Social Research, University of Michigan, Ann Arbor, MI, 1985, 133–166. This study investigated the total quantity of time American adults spend in one year performing various activities, distinguished by demographic characteristics, geographic location, and seasonal characteristics. The survey was based on data collected in the 1975–1976 Time Allocation Study for two weekdays, one Saturday, and one Sunday. Activity time estimates were made for 10 broad categories based on data collected from 87 activities (in response to the survey question "what were you doing?"). Data on time use for the major activity patterns were collected for four different age groups (18 to 24, 25 to 44, 45 to 64, and 65 and older). Two of the ten major categories were "house/yard work" and "personal care," presented in hours/week. Time estimates for these major activity categories were derived from several (of the 87) minor activity categories, including: "meal cleanup," "laundry," and "washing/dressing." The *Exposure Factors Handbook* (USEPA, 1997) provides values of the mean (hours/week) and standard deviation for each of these 87 activities specific to men, women, and men and women combined. Limitations to this study include that the data are somewhat dated, and that time use data for children are not presented. Advantages of this study include that the time diaries were not based on recall and that the survey is seasonally balanced.

- Robinson, J. P., and Thomas, J., Time Spent in Activities, Locations, and Microenvironments: A California-National Comparison: Project Report, Environmental Monitoring Systems Laboratory, U.S. Environmental Protection Agency, Las Vegas, NV, 1991. This study compared data from the 1987–1988 time activity study conducted by the California Air Resources Board (CARB) to data from the 1985 national study, American's Use of Time. In the CARB study, 1762 Californian adults (randomly sampled) completed a diary listing the previous day's activities and the location of each activity. In the national study, single day diaries were completed for over 5000 respondents across the U.S. aged 12 years and older. "Time Spent" data were collected for numerous activities, including: "meal cleanup," "clothes care," and "washing (personal)." "Time Spent" data were also collected for sixteen microenvironments, including "kitchen" and "bathroom." Limitations of this study include that the data were obtained through subject recall rather than through a real-time diary, that the subjects include only California residents, that distributional statistics were not provided for the data, and that the data represent only a single day for each subject. An advantage of the study is that it provides time estimates by activities, locations, and microenvironments grouped by age, gender, and type of day.

- California Air Resources Board (CARB), Study of Children's Activity Patterns, California Environmental Protection Agency, Air Resources Board Research Division, 1991. This study provided time estimates of 1200 children (11 years old or younger) in various activities and microenvironmental locations on a typical day. Data were acquired through the use of a 24-hour recall diary administered to either the child or the child's parent, depending on the child's age. The resulting data were classified according to 10 major activity groups, 113 detailed activities, 6 major categories of location, and 63 detailed microenvironmental locations. The detailed activity and microenvironmental information contain time estimates applicable to domestic water exposure assessments. Limitations of the study include the fact that the sample population was restricted to English-speaking children; the data were acquired through the use of a 24-hour recall diary; and the survey was conducted in California only. An advantage of this study is that it provides information on time expenditures in various activities and locations for children grouped by age, gender, and seasons.
- Tarshis, B., The "Average American" Book, New American Library, New York, N.Y., 1981, 191. The personal grooming data provided in this study are discussed above in Section 3.2.4.2 "Showering/Bathing."
- U.S. Environmental Protection Agency, Dermal Exposure Assessment: Principles and Applications, Report No. EPA/600/891/011B, Office of Health and Environmental Assessment, Washington, D.C., 1992. The range of default values for dermal exposure factors, including showering/bathing times, provided in this study is discussed above in Section 3.2.4.2, "Showering/Bathing."
- James, I. R., and Knuiman, M. W., An application of Bayes methodology to the analysis of diary records from a water use study, *Journal of the American Statistical Association,* 82(399), 705–711, 1987. The cumulative frequency distributions for average shower durations, provided by this study, are discussed above in Section 3.2.4.2, "Showering/Bathing."
- Robinson, J. P., and Blair, J., Estimating Exposure to Pollutants Through Human Activity Pattern Data: The National Micro-Environmental Activity Pattern Survey (MAPS), Survey Research Center, University of Maryland, College Park, MD, 1995. This research effort collected 24-hour time/activity data from 9386 people residing in the continental U.S. during the years 1992 to 1994. The study administered a recall diary, a follow-up questionnaire to identify specific pollutant sources, and a background questionnaire by telephone to each subject. Data were collected for use in estimating exposures relating to air quality, drinking water, pesticides, and toxins. The recall diary collected data concerning water consumption, use of various appliances (dishwashers, washing machines, humidifiers), cleaning activities, and bathing. Other questions inquired as to the subject's proximity to hot showers, door and window positions during bathing, and the household's source of drinking water (well, piped-in utility,

or bottled water). In the report, the typical study finding is presented as the percentage of a particular population group (e.g., Asian Americans) that engages in a specific water-related activity (e.g., consumption of bottled water). Although such data are useful for identifying the exposure potential of population groups, they are insufficient for an in-depth exposure assessment. The report by Tsang and Klepeis (1996) discussed below is a better source of data from the NHAPS study, as it provides frequency and duration statistics for a large number of water-related activities.

- Tsang, A. M., and Klepeis, N. E., Descriptive Statistics Tables for a Detailed Analysis of the National Human Activity Pattern Survey (NHAPS) Data, Report No. EPA/600/R-96/074, U.S. Environmental Protection Agency, July 1996. The authors conducted a detailed analysis of the NHAPS data and prepared an extensive set of tables listing frequency and duration statistics for a wide variety of data classifications. Useful tables include minutes per day spent in each room of the house (including bathroom, kitchen, and pool), minutes per day spent in selected nonresidential locations (including laundromat, pool/river/lake, and bar/restaurant), and minutes per day spent in selected activities (including food preparation, food cleanup, cleaning house, outdoor cleaning, clothes care, car repair/maintenance, plant care, animal care, other household work, washing, dishes/laundry, bathing, and eating/drinking). The report also provides data on humidifier use, number of loads of wash done while respondent was home, source of water (public system, private well, and other), gallons of bottled water used per week, glasses of tap water consumed, glasses of juice made with tap water, carbonated drinks consumed, number of hand washings, number of showers and baths, showers and baths with window or door open, swimming, dishwashing by hand vs. machine, clothes washing at home vs. laundromat, and location of home washing machine. In each case, the report provides statistics by gender, age, race, employment status, day of week, season, and disease group (asthma, angina, and bronchitis/emphysema).

The 1994 *Exposure Factors Sourcebook* (AIHC, 1994) notes the following study/resource:

- Johnson, T., McCoy, M., Capel, J. E., Wijnberg, L., and Ollison, W., A comparison of ten time/activity databases: effects of geographic location, temperature, demographic group, and diary recall method, in *Tropospheric Ozone: Nonattainment and Design Values Issues,* J. J. Vostal, Ed., Air and Waste Management Association, 1993. This study presents a comparison of ten time/activity surveys that were performed to increase understanding of how people move in and out of zones of varying air quality. Most of these 10 studies were conducted in California (e.g., the CARB study described above), but others were carried out in Cincinnati, OH, Denver, CO, Washington, D.C., and Valdez, Alaska. Recall-based diaries were used in some of these studies, while real-time diaries were used in others. The Cincinnati Activity Diary Study, performed by the Electric Power Research Institute (EPRI) in 1985, collected time/activity data for 2800 subject days and is the largest of the ten studies. Although these 10 studies were motivated by air quality concerns, the detailed activity pattern data provided in such studies would also be useful in estimating domestic water use patterns. For example, the Cincinnati study provides time/activity data concerning time spent in activities characterized as "meal preparation and clean-up" and "laundry."

The following study describes an innovative method for acquiring continuous water flow data specific to individual fixtures within the home.

- DeOreo, W.B., Hearney, J. P., and Mayer, P. W., Flow trace analysis to assess water use. *Journal of the AWWA,* January 1996. Data loggers equipped with magnetic sensors were fitted on residential water meters in 16 homes in Boulder, CO, between June and September, 1994. The loggers measured flow rates every 10 seconds to produce a continuous flow trace with sufficient precision to isolate, quantify, and categorize individual flow signatures for toilets, showers, bathtubs, dishwashers, clothes washers, irrigation, faucets, leaks, and swimming pools. The study found that toilets account for the majority of indoor water use, followed closely by clothes washers and showers. Miscellaneous faucet use and leaks are next; baths and dishwashers are almost insignificant. Data provided by this study may not be generally applicable; measurements represent a relatively small number of homes (16) in a single city (Boulder) during one season (summer).

Of the studies listed in this subsection, four studies appear to provide particularly useful and up-to-date data concerning water-use activities:

- Robinson, J. P., and Thomas, J., Time Spent in Activities, Locations, and Microenvironments: A California-National Comparison: Project Report, Environmental Monitoring Systems Laboratory, U.S. Environmental Protection Agency, Las Vegas, NV, 1991.
- California Air Resources Board (CARB), Study of Children's Activity Patterns, California Environmental Protection Agency, Air Resources Board Research Division, 1991.
- Robinson, J. P., and Blair, J., Estimating Exposure to Pollutants Through Human Activity Pattern Data: The National Micro-Environmental Activity Pattern Survey (MAPS), Survey Research Center, University of Maryland, College Park, MD, 1995.
- Tsang, A. M., and Klepeis, N. E., Descriptive Statistics Tables for a Detailed Analysis of the National Human Activity Pattern Survey (NHAPS) Data, Report No. EPA/600/R-96/074, U.S. Environmental Protection Agency, July 1996.

3.2.5 ONGOING STUDIES

The first stage of the National Human Exposure Assessment Survey (NHEXAS) is collecting questionnaire, microenvironmental, and biological data on a population-based sample of approximately 500 people in EPA Region V (Great Lakes Area) and Arizona. The study is also collecting questionnaire data from repeat visits to the same individual three times within a year in the Baltimore, MD area. Included in the baseline survey are questions soliciting information on demographic characteristics, housing characteristics, and personal activities. Data items include the presence of running water in the home; its source; water use in cooking and drinking; and any treatment, such as filtering. In the more detailed questionnaire administered during the environmental sampling are questions related to swimming pools, hot tub use, amounts and types of liquids consumed, and whether a bath or shower was taken on the day the questionnaire was administered. The field portion of this stage of NHEXAS is scheduled for completion in 1998.

A case control study of neural tube defects and drinking water contaminants was conducted by the New Jersey Department of Health and Human Services from 1995 to 1997. Questionnaires administered to several hundred women provide recall information on the amount of water and other beverages consumed, showers and baths taken, and the use of swimming pools and hot tubs. Technicians measured the water concentrations of trihalomethanes and haloacetic acids within each home at the time of the questionnaire data collection. A report summarizing the findings of this study is scheduled to be released by the Agency for Toxic Substance and Disease Registry in 1998.

3.2.6 DATA GAPS AND RESEARCH NEEDS

Currently, the researcher can find a variety of sources for data relating to (1) the frequency and times of activities producing *direct exposures*, (2) the duration of these activities, and (3) the quantity of water used during each activity. More data are needed to adequately characterize the *indirect exposures* that occur when one household member is exposed through the water- related activities of another household member. Time/activity studies should be conducted in which the relative locations of all household members are monitored throughout the time period of interest. The resulting "household" databases could be used to develop occurrence probabilities for indirect exposure events such as one household member entering a bathroom while another member is showering.

3.2.7 PRIMARY REFERENCES

These references provide useful summaries and evaluations of many of the individual citations presented in the section.

- U.S. Environmental Protection Agency, The *Exposure Factors Handbook,* EPA/600/P-95/002Fc, Washington, D.C., 1997.
- American Industrial Health Council (AIHC), *Exposure Factors Sourcebook,* Washington, D.C., May 1994.

- U.S. Department of Commerce, *1990 Census Questionnaires and Other Public-Use Forms,* Report No. 1990 CPH-R-5, Bureau of the Census, Washington, D.C., 1990.
- U.S. Environmental Protection Agency, Methodology for Assessing Health Risks Associated with Indirect Exposure to Combustor Emissions (Interim Final Report), Report No. EPA/600/6-90/003, Office of Health and Environmental Assessment, Washington, D.C., 1990.

REFERENCES

American Industrial Health Council (AIHC), *Exposure Factors Sourcebook,* Washington, D.C., May 1994.

Brown, H. S., Bishop, D. R., and Rowan, C. A., The role of skin absorption as a route of exposure for volatile organic compounds (VOCs) in drinking water, *Am. J. Publ. Health,* 74(5), 479-484, May, 1984.

California Air Resources Board (CARB), Study of Children's Activity Patterns, California Environmental Protection Agency, Air Resources Board Research Division, 1991.

Canadian Ministry of National Health and Welfare, Tap water Consumption in Canada, Document No. 82-EHD-80, Public Affairs Directorate, Department of National Health and Welfare, Ottawa, Canada, 1981.

Cantor, K. P., Hoover, R., Hartge, P., Mason, T. J., Silverman, D. T. et al., Bladder cancer, drinking water source, and tap water consumption: a case-control study, *J. Natl. Cancer Inst.,* 79(6), 1269–1279, 1987.

DeOreo, W. B., Hearney, J. P., and Mayer, P. W., Flow trace analysis to assess water use. *J. AWWA,* 79–90, January 1996.

East Bay Municipal Utility Water District, Urban Water Management Plan, written communication to J. B. Andelman, July 1992.

Ershow, A. G., Brown, L. M., Cantor, K. P., Intake of tapwater and total water by pregnant and lactating women, *Am. J. Publ. Health,* 81, 328–334, 1991.

Ershow, A. G., and Cantor, K. P., Total Water and Tapwater Intake in the United States: Population-Based Estimates of Quantities and Sources, Life Sciences Research Office, Federation of American Societies for Experimental Biology, 1989.

Gillies, M. E., and Paulin, H. V., Variability of mineral intakes from drinking water: a possible explanation for the controversy over the relationship of water quality to cardiovascular disease, *Int. J. Epid.,* 12(1), 45–50, 1983.

Hakkinen, P. J., Cleaning and laundry products, human exposure assessments, in *Handbook of Hazardous Materials,* Korn, M., Ed., Academic Press, San Diego, 145-151, 1993.

Hill, M. S., Patterns of time use, in *Time, Goods, and Well-Being,* Juster, F. T., and Stafford, F. P., Eds., Survey Research Center, Institute for Social Research, University of Michigan, Ann Arbor, MI, 133–166, 1985.

Hopkins, S. M., and Ellis, J. C., Drinking Water Consumption in Great Britain: A Survey of Drinking Habits with Special Reference to Tap-Water-Based Beverages, Technical Report No. 137, Water Research Centre, Wiltshire, Great Britain, 1980.

International Commission on Radiological Protection (ICRP), *Report of the Task Group on Reference Man,* Pergamon Press, New York, 1981.

Israeli, M., and Nelson, C. B., Distribution and expected time of residence for U.S. households, *Risk Anal.,* 12(1), 65–72, 1992.

James, I. R., and Knuiman, M. W., An application of Bayes methodology to the analysis of diary records from a water use study, *J. Am. Stat. Assoc.*, 82(399), 705–711, 1987.

Johnson, T., McCoy, M., Capel, J. E., Wijnberg, L., and Ollison, W., A comparison of ten time/activity databases: effects of geographic location, temperature, demographic group, and diary recall method, in *Tropospheric Ozone: Nonattainment and Design Values Issues*, J. J. Vostal, Ed., Air and Waste Management Association, 1993.

Johnson, T., and Capel, J., A Monte Carlo Approach to Simulating Residential Occupancy Periods and Its Application to the General U.S. Population, Report No. 450/3-92-011, U.S. Environmental Protection Agency, Office of Air Quality and Standards, Research Triangle Park, N.C., 1992.

Juster, F. T., Hill, M. S., Stafford, F. P., and Parsons, J. E., Study Description: 1975–1981 Time Use Longitudinal Panel Study, Survey Research Center, Institute for Social Research, University of Michigan, Ann Arbor, MI, 1983.

McNall, P. E., and Schlegel, J. C., Practical thermal environmental limits for young adult males working in hot, humid environments, *American Society of Heating, Refrigerating and Air-Conditioning Engineers (ASHRAE) Transactions*, 74, 225–235, 1968

Metropolitan Water District of Southern California, Urban water use characteristics in the metropolitan water district of Southern California. Draft Report, August 1991.

Murray, D. M., Residential house and zone volumes in the United States: empirical and estimated parametric distributions, *Risk Anal.*, 17 (4), 439-446, 1997.

National Academy of Sciences (NAS), *Drinking Water and Health, Vol. 1*, National Academy of Sciences–National Research Council, Washington, D.C., 1977.

Nazaroff, W. W., and Nero, A. V. (Eds.), *Radon and Its Decay Products in Indoor Air*, John Wiley & Sons, New York, 1988.

Pennington, J. A. T., Revision of the total diet study food list and diets, *J. Am. Diet. Assoc.*, 82, 166–173, 1983.

Robinson, J. P., Changes in Americans' Use of Time: 1965–1975 — A Progress Report, Cleveland State University: Communication Research Center, Cleveland, Ohio, 1977.

Robinson, J. P., and Blair, J., Estimating Exposure to Pollutants Through Human Activity Pattern Data: The National Micro-Environmental Activity Pattern Survey (MAPS), Survey Research Center, University of Maryland, College Park, MD, 1995.

Robinson, J. P., and Thomas, J., Time Spent in Activities, Locations, and Microenvironments: A California-National Comparison: Project Report, Environmental Monitoring Systems Laboratory, U.S. Environmental Protection Agency, Las Vegas, NV, 1991.

Rogan, S., "Utilizing Decennial Census Data and the Value of 'Field Truthing' in Exposure Assessment Research," 7th Annual Meeting of the International Society of Exposure Analysis, 1997.

Roseberry, A. M., and Burmaster, D. E., Lognormal distributions for water intake by children and adults, *Risk Anal.*, 12, 99–104, 1992.

Sedman, R., Funk, L., and Fountain, R., Distribution of residence duration in owner-occupied housing, *J. Expos. Anal. Environ. Epidemiol.*, 8 (1): 51–58, 1998.

Sivadas, E., Matthew, G., and Curry, D. J., A preliminary examination of the continuing significance of social class to marketing: a geodemographic replication, *Journal of Consumer Marketing*, 14, 463–479, 1997.

Tarshis, B., The "Average American" Book, New American Library, New York, 1981, 191.

Timmer, S. G., Eccles, J., and O'Brien, K., How children use time, in *Time, Goods, and Well-Being*, Juster, F. T., and Stafford, F. P., Eds., Survey Research Center, Institute for Social Research, University of Michigan, Ann Arbor, MI, 353–380, 1985.

Tsang, A. M., and Klepeis, N. E., Descriptive Statistics Tables for a Detailed Analysis of the National Human Activity Pattern Survey (NHAPS) Data, Report No. EPA/600/R-96/074, U.S. Environmental Protection Agency, July 1996.

U.S. Army, Water Consumption Planning Factors Study, Directorate of Combat Developments, United States Army Quartermaster School, Fort Lee, Virginia, 1983.

U.S. Bureau of Census, *Geographical Mobility: March 1991 to March 1992, Current Population Reports,* p. 20–473, 1993.

U.S. Bureau of Census, *Statistical Abstract of the United States: 1985 (105th ed.),* U.S. Government Printing Office, Washington, D.C., 1984.

U.S. Bureau of Census, *Statistical Abstract of the United States: 1992 (112th edition),* Washington, D.C., 1992, Table 1230, p. 721.

U.S. Department of Commerce, *1990 Census Questionnaires and Other Public-Use Forms,* Report No. 1990 CPH-R-5, Bureau of the Census, Washington, D.C., 1990.

U.S. Department of Energy, *Housing Characteristics 1990,* Report No. DOE/EIA-0314 (90), Energy Information Administration, Washington, D.C., 1992.

U.S. Department of Housing and Urban Development, Residential Water Conservation Projects: Summary Report, Report No. HUD-PDR-903, Office of Policy Development and Research, Washington, D.C., 1984.

U.S. Environmental Protection Agency, An Estimation of the Daily Average Food Intake by Age and Sex for Use in Assessing the Radionuclide Intake of Individuals in the General Population, Report No. EPA-520/1-84-021, 1984.

U.S. Environmental Protection Agency, Dermal Exposure Assessment: Principles and Applications, Report No. EPA/600/891/011B, Office of Health and Environmental Assessment, Washington, D.C., 1992.

U.S. Environmental Protection Agency, The Exposure Factors Handbook, Office of Health and Environmental Assessment, Washington, D.C., 1989.

U.S. Environmental Protection Agency, The Exposure Factors Handbook, EPA/600/P-95/002Fc, Washington, D.C., 1997.

U.S. Environmental Protection Agency, Methodology for Assessing Health Risks Associated with Indirect Exposure to Combustor Emissions (Interim Final Report), Report No. EPA/600/6-90/003, Office of Health and Environmental Assessment, Washington, D.C., 1990.

U.S. Environmental Protection Agency, Superfund Exposure Assessment Manual, Report No. EPA/540/1-88/001, NTIS PB89-167985, Office of Emergency and Remedial Response, Washington, D.C., 1988.

U.S. Environmental Protection Agency, Water quality criteria documents; availability, *Federal Register,* 45(231), 79318–79379 (November 28, 1980).

Wilkes, C. R., Small, M. J., Davidson, C. I., and Andelman, J. B., The effects of behavior on human inhalation exposure to contaminants volatilized during indoor water uses, *J. Expos. Anal. Environ. Epidemiol.,* 6 (4), 393–412, 1996.

3.3   BUILDING CHARACTERISTICS

P. J. (Bert) Hakkinen

As noted in other parts of this document, modeling of potential exposures to indoor volatile chemicals requires information about certain key characteristics of the residence. There are four types of input data generally required for indoor modeling:

"(1) house and room volumes, (2) residence times for air in each household volume, (3) water use by category, and (4) amount of time individuals spend in the shower, bathroom, and remaining house..." (McKone, 1989). The major purpose of Section 3.3 is to provide a critical analysis of data available for categories (1) and (2) needed for indoor air modeling, with categories (3) and (4) addressed elsewhere in this chapter. Thus Section 3.3 addresses:

- The volumes of residences and residential zones and rooms.
- The ventilation characteristics (e.g., air changeovers per hour) of residences and residential zones and rooms.

The key reference is U.S. EPA's *Exposure Factors Handbook*. The sections and table numbers noted in Section 3.3 are from the latest edition, (published in 1997), with the entire text and tables also available via the Internet's World Wide Web (http://www.epa.gov/ORD/WebPubs/exposure), and as a CD-ROM (information about these different forms of the *Exposure Factors Handbook* is available via the Internet from moya.jacqueline@epamail.epa.gov). The latest edition is the first revision since the first edition was published in 1988–1989, and its Chapter 16, "Reference Residence," summarizes key information about building characteristics.

Noteworthy is that the contents of Chapter 16 and the rest of the revised *Exposure Factors Handbook* are the result of a several-year effort involving extensive information searches, peer review workshops, and other reviews to help ensure that the *Handbook* contains the most relevant and useful information. The ILSI workgroup, whose membership includes several authors, contributors, and reviewers of the *Exposure Factors Handbook*, concurs with the studies and data chosen for inclusion in Chapter 16 of the *Exposure Factors Handbook*. Chapter 16 of the *Exposure Factors Handbook* is organized as follows, with key information from Sections 16.2, 16.3, and 16.6 noted in the rest of Section 3.3:

While Chapter 16 of the *Exposure Factors Handbook* should be the primary source of information about building characteristics, the following publications provide perspective beyond the *Exposure Factors Handbook* via case studies showing why specific residential factors are needed:

- Anonymous, 1994.
  Key information included: Pages 29–30 and 35–39 of this report from the European Centre for Ecotoxicology and Toxicology of Chemicals demonstrate the use of house and room volume and air exchange rate data, and the use of EPA's Screening-Level Consumer Inhalation Exposure Software (SCIES) in the assessment of inhalation exposure to volatile consumer product components.
- Finley et al., 1993, McKone and Bogen, 1992, and Wilkes et al., 1992.
  Demonstrates the use of house, shower, and bathroom air exchange rates; shower and house water use rates; shower, bathroom, and house exposure times, and transfer efficiencies from water to shower air and household air.
- Wilkes et al.,1996
  Models the impact of an individual's activities and those of other household occupants on inhalation exposures to volatiles from household water. Includes use of time/activity pattern data for household occupants, along with house, shower, bathroom and other air exchange rates, shower and other water use rates, etc.

3.3.1   RESIDENCE TYPES AND SUBTYPES, AND THEIR VOLUMES

Many types of U.S. residences and residential designs exist and are of potential interest to exposure and risk assessors. Residence types include single-family attached or detached dwellings, multifamily units, mobile homes, etc. Subtypes include ranch houses, two-story houses, etc. Given the impact that residence type can have on residential exposure parameter values, the ability to have house and room volume information for various residence types and subtypes is crucial to residential exposure assessments.

Section 16.2 of the *Exposure Factors Handbook* contains valuable information on U.S. building characteristics, including the following:

- Table 16.1 summarizes distributions of U.S. residential volumes (from Thompson, 1995 and Versar, 1990). For example, the arithmetic mean of the residential volume distributions in both studies is 369 m³.
- Table 16.2 provides U.S. residential volumes in relation to housing type (single-family detached, single-family attached, multifamily 2 to 4 units, multifamily 5+ units, mobile home, and "all types") and ownership (owner-occupied, rental, and "all units") (from U.S. DOE, 1995). For example, the average estimated volume of a single-family detached, owner-occupied residence is 471 m³.
- Table 16.3 provides U.S. residential volumes in relation to household size (1, 2, 3, 4, or 5 persons, 6 or more persons, and "all sizes") and year of construction (1939 or before, 1940–1949, 1950–1959, etc., and "all years") (from U.S. DOE, 1995). As examples, the volume of a four-person residence is 431 m³, while an "all sizes" residence is 369 m³. Also, a residence built during 1970 to 1979 had a volume of 350 m³, while the volume of a residence built "all years" is 369 m³.

Section 16.6 (Recommendations) of the *Exposure Factors Handbook* recommends that a mean value of 369 m³ (the arithmetic mean of the residential volume distributions in both studies in Table 16.1) be used as a central estimate of the residential volume, and that a value of 217 m³ (the average of the 25th percentile values of both studies in Table 16.1) be used for developing conservative estimates of exposure.

Section 16.2.2 of the *Exposure Factors Handbook* provides information on volumes of residential rooms, including data from a recent study (Murray, 1997; called "Murray, 1996" in Chapter 16 of the *Exposure Factors Handbook*). The Murray publication contains distributions of volumes for the following areas of residences (the text of Section 16.2.2 only provides the means and standard deviations):

- Basements
- First floors
- Second floors
- Kitchen zones (kitchen, plus any of the following: utility room, dining room, living room, or family room)
- Bedroom zones (all of the bedrooms, plus bathrooms and any halls associated with the bathrooms)

As further information useful for some exposure assessments, Section 16.2.2 of the *Exposure Factors Handbook* provides information about surface areas of walls, floors, and ceilings, and the types and amounts of products and materials used in constructing and finishing indoor surfaces. Table 16.5 provides product and material examples useful as typical "amount of surface covered" values for a residence. For example, floor wax is assumed to cover 50.0 m^2, and wallpaper is assumed to cover 100.0 m^2.

3.3.2 AIR EXCHANGE RATES

Ventilation can also be called "airflow rates," "air exchange rates," and "air changeovers per hour," and will be called air exchange rates for the rest of this section. Until recently, air exchange rate data for U.S. residences were scattered among numerous publications (e.g., journal articles and government and contractor reports). Fortunately, there have been several recent attempts to compile and review information and data from thousands of residences. The key data are presented in Section 16.3.2 of the *Exposure Factors Handbook* as Tables 16.8, 16.9, and 16.10, and include air exchange distributions as a function of region of the U.S. climatic region, and season of the year (Versar, 1990, Koontz and Rector, 1995, and Murray and Burmaster, 1995).

Section 16.6 (Recommendations) of the *Exposure Factors Handbook* recommends that, based on the results shown in Table 16.9, a median value of 0.45 air changes per hour be used as a typical value (this value is noted by EPA as being "very close to the geometric mean of the measurements" from the large number of air exchange rate studies shown in Table 16.8 and summarized in Table 16.9). A value of 0.18 air changes per hour is recommended in the *Exposure Factors Handbook* for developing conservative estimates of exposure from indoor sources (this is the 10th percentile value from the large number of air exchange studies summarized in Table 16.10).

One air exchange publication not included in the *Exposure Factors Handbook* is Pandian et al., 1993. This publication includes data from numerous studies and generates frequency distributions and summary statistics for residential air changeovers in different regions of the United States, different seasons, and for homes with single vs. multiple levels. However, since the 1993 publication, there has been acknowledgment of errors in some of the initial data sets used. These errors have been accounted for and corrected where possible in the publication by Murray and Burmaster, 1995 (Table 16.10 of the *Exposure Factors Handbook*).

3.3.3 AIRFLOW BETWEEN ROOMS AND ZONES

The *Exposure Factors Handbook* (Section 16.3.5, Interzonal Airflow) does not recommend any values to use for airflow between rooms and zones, noting that:

> Existing data represent a variety of case studies, but do not point to any general rules relating to directional airflows. Because existing data represent a variety of case studies, the data are inadequate for defining general characteristics that represent typical or average conditions. Further, the definition of airflow zones in residential structures can be ambiguous because occupants open and close interior doors on irregular schedules.

If room and zone airflow data are judged to be required for an exposure assessment, the ILSI workgroup notes that Finley et al., 1993 and McKone and Bogen, 1992 provide the same probability density functions for shower and bathroom air exchange rates, which are based on various assumptions by McKone about the volume and range of air changes per hour of a shower and bathroom. Other air exchange rate information for residential rooms and zones are available, e.g., measured bathroom ventilation rates as reported in Wilkes et al., 1996; however, the information in Finley et al., 1993 and McKone and Bogen, 1992 is currently suggested for exposure assessors requiring values to use for shower and bathroom air exchange rates.

3.3.4 Key Factors Affecting Degree of Air Exchange and Airflow (seasonal and geographic/climatic influences, opening of windows and doors, and use of mechanical air handling systems)

The studies noted above and in the *Exposure Factors Handbook* contain air exchange rate data on U.S. homes in various geographic regions and climates, and for different seasons. For example, one study (Murray and Burmaster, 1995) contains air exchange rate data for different climatic regions and for different seasons. The climatic regions were based on "heating degree days," which account for indoor/outdoor temperature differences (see Section 16.3.2 and Table 16.10 of the *Exposure Factors Handbook*). An example of the impact of climatic region is that the median air exchange rate varied up to about twofold depending on the season being compared. An example of the impact of season is that the median air exchange rate within a climatic region varied up to several-fold depending on the region and season. As might be expected, the highest air exchange rates tended to occur in the warmest climates.

A recent study found that opening exterior windows and doors was the most important determinant of air exchange rate variation in the one house studied (Wallace and Ott, 1996), with a maximum air exchange rate of about 12 air changes per hour observed when all doors and windows were open compared to the house's typical air exchange rate of about 0.4. This study also examined the impact of opening one or more windows different amounts, closing off some rooms, the use of air conditioning, and indoor-outdoor temperature differences.

Other factors that affect air exchange rates include the following that could be considered for assessments of specific residences (Sections 16.2.3, 16.2.4, and 16.3.1 of the *Exposure Factors Handbook* and Pandian et al., 1993 contain further discussion and citations of publications addressing these factors. Also, Pandian's "Residential Exposure Factors" chapter in the U.S. EPA/International Society of Exposure Analysis/Society for Risk Analysis *Residential Exposure Assessment: A Source Book* will be a very useful resource; more information about this book is provided in Section 3.3.5 below):

- The natural "tightness" of the residence, e.g., newer residences in colder regions have heavier insulation, vapor barriers, and weather stripping, which tend to lower air exchange rates.
- The intentional (designed) intake, distribution, and removal of air through ventilation systems, e.g., central, localized, or portable heating and air conditioning, localized exhaust, and/or systems designed to increase overall air exchange rate in a residence. As noted in Section 16.2.3 of the *Exposure Factors Handbook,* continuous operation of an airflow system rated at 100 ft^3/min (or 170 m^3/h) could increase the air exchange rate in a 4000 m^3 house by about 0.4 air changes per hour.

- Strong winds.
- Structural channels such as chimneys, crawl spaces, and stairwells.

3.3.5 APPLICABLE REFERENCES AND DATABASES

As emphasized by the above discussion, the key "all-in-one" source for residential exposure factor data is the latest version of U.S. EPA's revised *Exposure Factors Handbook*. In addition, the U.S. EPA, the International Society of Exposure Analysis, and the Society for Risk Analysis are developing the *Residential Exposure Assessment: A Source Book,* with the expectation that it will be an excellent guidance and training document.

Residential Exposure Assessment: A Source Book, to be published in 1999, is being designed to serve as the key overall source of guidance and training on assessment of residential exposures. As such, it will cite information in the revised *Exposure Factors Handbook* while attempting to minimize overlap in content between the two documents. This book will include an overview of the principles and practices of residential exposure assessment, general considerations for residential exposure assessment, evaluating exposures from indoor and outdoor use consumer products, methodologies for assessing dermal, inhalation, and oral exposures, and key human and residential exposure factors. Of particular interest to readers of this chapter will be the "Residential Exposure Factors: An Overview" chapter containing the following sections:

- Introduction to residential exposure factors, including human activity patterns, residential environments, and sources of contaminants
- Models for human activity patterns, (indoor) environments, and contaminant sources
- Housing stock, including geographic distribution, building materials, and appliances
- Sources of contaminants — locations, use, and release rates
- Ventilation parameters — air exchange rate, air recirculation rate, interzonal air flows, intrazonal circulation
- Effective volumes
- Surface characteristics — type and area
- Water uses

3.3.6 DATA GAPS AND RESEARCH NEEDS FOR BUILDING CHARACTERISTICS

The most thorough recent discussion and summary of residential exposure factor data gaps and research needs is judged to be from a July, 1995, EPA workshop held to help finalize the revised *Exposure Factors Handbook*. The "Housing Characteristics and Indoor Environments" work group at this 1995 workshop identified the following exposure factor data gaps and research needs:

- Source emissions as having the most importance
  (Note: The following are not listed in order of recommended importance.)
- Urban vs. suburban vs. rural housing characteristics

- Ceiling height as a function of the year of construction and location (urban vs. suburban vs. rural)
- Single vs. multifamily residences, including representative building plans
- Ensuring all useful air exchange data sets are covered
- Appliance characteristics (temperatures and volumes)
- Building materials and furnishings
- Mechanical system configurations and rates (includes information on kitchen, bathroom, and newer mechanical ventilation systems)
- Transport (e.g., soil tracking)
- Need for reality checks of exposure assessments, including factor values used, and the need for validation of models

Several of the above data gaps have been addressed since the 1995 EPA workshop. For example, Murray, 1997 (cited in the *Exposure Factors Handbook* as Murray, 1996) includes house volume data for housing types noted above as data gaps, e.g., single-family and multifamily, and also as a function of ownership status (own or rent) and urban status (central city, suburban, or rural). Also, EPA's *Exposure Factors Handbook* contains sections discussing several of these data gaps and noting available useful data, e.g., products and materials used in constructing or finishing residential surfaces are discussed in Section 16.2.2, mechanical system configurations are discussed in Section 16.2.3, and appliance temperatures and volumes are discussed in Section 16.3.6.

3.3.7 ADDITIONAL PUBLICATION

As noted elsewhere in this document (e.g., Sections 3.2.3 and 3.2.4.4), a recent activity pattern survey for the U.S. EPA of over 9000 people provides information on many aspects of residential living and activities, including various indoor tasks (e.g., taking showers or baths, if a window was open or a fan was on during the bath, etc.), the use of various appliances (e.g., dishwashers, washing machines, and humidifiers), etc. [Robinson and Blair, 1995; cited in the *Exposure Factors Handbook*'s Chapter 14 ("Activity Factors," Tables 14.19 to 14.139) as Tsang and Klepeis, 1996]. While this survey is noted by the ILSI workgroup as a key source of perspective and data, the survey did not gather information on key residential characteristics such as house and room volumes and ventilation (air exchange rates).

REFERENCES

Anonymous, *Technical Report No. 58. Assessment of Non-Occupational Exposure to Chemicals,* European Centre for Ecotoxicology and Toxicology of Chemicals, Brussels, Belgium, 1994, 29–30 and 35–39.

Finley, B. L., Scott, S., and Paustenbach, D. J., Evaluating the adequacy of maximum contaminant levels as health-protective cleanup goals: an analysis based on Monte Carlo techniques, *Regul. Toxicol. Pharmacol.,* 18: 438, 1993.

Koontz, M. D. and Rector, H. E., Estimation of distributions for residential air exchange rates, EPA Contract No. 68-D9-0166, work assignment No. 3-19, U.S. EPA, Office of Pollution Prevention and Toxics, Washington, D.C., 1995 (as cited in Section 16.3.2 of the U.S. EPA *Exposure Factors Handbook* noted below).

McKone, T. E., Household exposure models, *Toxicol. Lett.*, 49, 321, 1989.

McKone, T. E. and Bogen, K. T., Uncertainties in health-risk assessment: an integrated case study based on tetrachloroethylene in California groundwater, *Regul. Toxicol. Pharmacol.*, 15, 86, 1992.

Murray, D. M., Residential total house and zone volumes in the United States: empirical and estimated parametric distributions, *Risk Anal.*, 17, 439, 1997 (cited as Murray 1996 in Section 16.2.2 of the U.S. EPA *Exposure Factors Handbook* noted below).

Murray, D. M. and Burmaster, D. E., Residential air exchange rates in the United States: empirical and estimated parametric distributions by season and climatic region, *Risk Anal.*, 15, 459, 1995.

Pandian, M., Ott, W. R., and Behar, J. V., Residential air exchange rates for use in indoor air and exposure modeling studies, *J. Expos. Anal. Environ. Epidemiol.*, 3, 407, 1993.

Robinson, J. P. and Blair, J., Draft of estimating exposure to pollutants through human activity pattern data. The national micro-environmental activity pattern survey (MAPS). Prepared for U.S. EPA, 1995 (as cited in Section 16.3.7 of the U.S. EPA *Exposure Factors Handbook* noted below).

Thompson, W., U.S. Department of Energy and Energy Information Administration, Personal communication on distribution of heated floor space area from the 1993 RECS, December 1995 (as cited in Section 16.2.1 of the U.S. EPA Exposure Factors Handbook noted below).

Tsang, A. M. and Klepeis, N. E., Results tables from a detailed analysis of the National Human Activity Pattern Survey (NHAPS) response. Draft report prepared for the U.S. Environmental Protection Agency by Lockheed Martin, Contract No. 68-W6-001, Delivery Order No. 13, 1996 (as cited in Section 14.1.1 of the U.S. EPA *Exposure Factors Handbook* noted below).

U.S. Department of Energy (DOE), Housing characteristics 1995, Report No. DOE/E1A-0314 (93), Energy Information Administration, Washington, D.C., 1995 (as cited in Section 16.2.1 of the U.S. EPA *Exposure Factors Handbook* noted below).

U.S. Environmental Protection Agency (EPA), Chapter 16, Reference residence, In Office of Research and Development, National Center for Environmental Assessment, *Exposure Factors Handbook,* EPA/600/P-95/002Fc, Volume 3 of 3. U.S. EPA, Washington, D.C., pp 16-1–16-33, 1997. *The entire text and tables are available via the Internet's World Wide Web at http://www.epa.gov/ORD/WebPubs/exposure, and as a CD-ROM (information about these different forms of the Exposure Factors Handbook is available via the Internet from moya.jacqueline@epamail.epa.gov).*

U.S. EPA, International Society of Exposure Analysis, and Society for Risk Analysis, Residential exposure assessment: a source book, Plenum Press, New York, 1999.

Versar, Database of PFT ventilation measurements: Description and user's manual, U.S. EPA contract No. 68-0204254, Task No. 39, U.S. EPA, Office of Toxic Substances, Washington, D.C., 1990 (as cited in Section 16.3.2 of the U.S. EPA *Exposure Factors Handbook* noted above).

Wallace, L. A. and Ott, W. R., Air exchange rate measurements in a detached house using a continuous monitor, abstract #A10.02, Joint 1996 annual meeting of the Society for Risk Analysis and International Society of Exposure Analysis.

Wilkes, C. R., Small, M.J., Andelman, J. B. et al., Inhalation exposure model for volatile chemicals from indoor uses of water, *Atmos. Environ.*, 26A, 2227, 1992.

Wilkes, C. R., Small, M. J., Davidson, C. I. et al., Modeling the effects of water usage and co-behavior on inhalation exposures to contaminants volatilized from household water, *J. Expos. Anal. Environ. Epidemiol.*, 6, 393, 1996.

3.4 WATER SOURCE CHARACTERISTICS

David A. Reckhow

There is a great variety of contaminant concentrations in U.S. drinking waters. The reasons for this variability are related to the source water quality as well as the physical, chemical, and biological environment that the water is exposed to prior to use by the consumer. Each of these contributing or mitigating factors will be examined in this chapter, as will the current state of knowledge of actual contaminant concentrations.

3.4.1 ANALYSIS OF SYSTEM COMPONENTS

The water that a consumer is exposed to will differ for different regions in the U.S., and even for different households in a single community. Furthermore, this water quality will change with the seasons. The types and concentrations of contaminants that are present in tap water are a product of the raw water source and all processes and environments to which the water is exposed after it enters a water system.

3.4.1.1 Nature of Supply

Perhaps the most important single characteristic of a water supply is whether it is from the surface or subsurface. About 20% of all community water systems in the U.S. use surface water; the remaining 80% use groundwater. However, the surface water systems tend to be much larger, so that the population served by surface water sources is about two thirds of the total.

The natural matrix of a water is determined by its geographic location and the surrounding geology and flora. Processes that contribute to the natural hardness, alkalinity, and salinity of a water are reasonably well understood (e.g., Stumm & Morgan, 1996). Somewhat less is known about the origin of natural organic matter (NOM). The NOM is especially important because it can react with oxidants and disinfectants in drinking water systems, forming toxic organic compounds. It will also elevate the solubility of certain hydrophobic organic contaminants.

Processes that contribute to NOM levels in natural waters are quite complex and locally variable. For this reason, it is not practical to try to predict NOM levels for a certain locale. Rather, they should be determined by direct measurement. Likewise, the levels of anthropogenic pollutants must be determined for each specific water body and groundwater well. It may be possible to project likely contamination in a susceptible aquifer; however, the location of a specific contaminant plume must be determined empirically.

3.4.1.2 Centralized Treatment

Community water systems serve about 83% of the total U.S. population. Most of these employ some form of treatment to make the water microbiologically and chemically safe. The impact of these treatment systems can profoundly change the composition of a water. Many of these changes are well understood, and mathematical models have been developed to predict the effects of various treatment scenarios (e.g., Black et al.,1996).

Almost all community water systems use a chemical disinfectant. Most use some form of hypochlorite. Many plants also coagulate and filter their water, and the number of unfiltered systems is expected to dwindle in the coming years. It is also becoming more common to use alternative preoxidants like ozone and permanganate. Each of these processes will remove selected groups of raw water contaminants, some will add new ones. The degree to which this occurs depends on a range of engineering factors (e.g., chemical doses, reaction times), environmental factors (e.g., temperature), and raw water quality.

To help the water treatment engineers select the best conditions for their plants, a series of mathematical models have been developed. Perhaps the most comprehensive of these treatment plant models is the one developed by James Montgomery, Inc. and Malcolm Pirnie, Inc. under contract through the Water Industry Technical Action Fund (WITAF) and the AWWA (WITAF, 1993). The objective of this modeling effort was to predict disinfection byproduct concentrations in treated waters. It encompasses submodels for removal of NOM by alum coagulation, ferric coagulation, lime softening, activated carbon adsorption, and membrane filtration. In addition, there are models for disinfection byproduct formation following final disinfection with chlorine.

Another group of mathematical models has been developed around removal of specific raw water contaminants by a specific process, usually activated carbon adsorption or membrane filtration. These are well developed, and generally work quite well for waters low in organic matter. However, they are less successful in predicting the impacts of NOM on long-term adsorption characteristics or on membrane fouling.

After leaving a treatment plant, the water must pass through distribution pipes to reach the consumer. During this passage, water quality can change appreciably. Disinfection byproducts can continue to form and decay. In addition, corrosion can occur, leading to the release of corrosion byproducts (i.e., metals and metal oxides). Water quality models are currently being developed for use with pipe distribution systems. Prominent among these are the EPANET (USEPA) and STONER (Severn Trent Water) programs. These efforts are still quite new, and much refinement is needed. Special concerns and needs include better knowledge of disinfection reactions at pipe walls and better understanding of the chemistry of corrosion.

3.4.1.3   Point of Use Treatment

A point of use (POU) water system may be installed just after the entry of a service line so that it treats all water used at a specific location (e.g., a house). More commonly, it may be installed at a single tap or at some intermediate location. If well designed and well operated, point of use treatment systems can be very effective at removing drinking water contaminants. However, even the best system will fail if not maintained. Failure of adsorption systems can result in "chromatographic" desorption of contaminants and an actual increase in their concentrations over that which is entering the building.

Very little is known about the actual performance of POU systems. Although, the number of POU systems is still relatively small, this is an area that needs more attention from researchers.

3.4.1.4 Impact of Home Plumbing and Appliances

Very little is known about the impact of home plumbing and appliances on drinking water quality characteristics. Some chemical processes are well known, such as precipitation of calcium carbonate in water heaters and dissolution of metals (e.g., copper, iron, lead) in pipes and plumbing fixtures. However, there are a number of other possible effects that have not been fully examined. These might include temperature effects on the decomposition of oxidant residuals, organic oxidation byproducts, and reaction of oxidant residuals with specific trace contaminants. The presence of certain additives (e.g., surfactants and chelators in detergents, soaps, water conditioners) in home appliances may also alter the physical and chemical properties of the water. In an effort to address some of these questions, Weisel and Chen (1994) have recently shown that some chlorination byproducts increase in concentration and others decrease when tap water is heated to 65%C.

Home appliances and fixtures also represent a potential source of contaminants. Aside from the well-known corrosion byproducts, there are materials in appliances that might leach out organic compounds. These may be plastics or rubber components that contain leachable monomers, or paints and coatings containing solvents or other constituents that can be released into the water. Finally, the possibility of residual oxidants reacting with materials in appliances and releasing new oxidation byproducts into the water must be considered.

3.4.2 ESTIMATION OF SOURCE WATER QUALITY

The estimation of source water quality must be undertaken from an empirical point of departure. Even with a perfect water treatment and distribution model, one would still need information on specific raw water quality to calibrate the model. In actuality, the perfect treatment and distribution models do not yet exist. This requires that a mostly or entirely stochastic approach be taken.

In examining source water quality, a decision must be made as to the desired cohort or geographic resolution. In this subsection, available data and methods for making estimates on different geographical scales are discussed.

3.4.2.1 Nationwide Estimates

At present there are no complete compilations of the nation's drinking water data. This type of information appears in a fragmented or incomplete form in a variety of published and unpublished reports (see Table 3.4.1).

The nation's 59,000 community water suppliers are required by their State Departments of Health to monitor product water quality for a variety of contaminants. Very little of this information has been compiled or summarized. Individual water systems (e.g., private wells) are generally not tested for chemical contaminants.

TABLE 3.4.1
Some Important National Drinking Water Quality Surveys

Survey	Organization	Sampling Dates	Data Collected	Scope	Citation
National Organics Reconnaissance Survey (NORS)	USEPA	1975	THMs in all; up to 129 organics in a subset of 10 cities	80 U.S. cities	Symons et al. 1975
National Organics Monitoring Survey (NOMS)	USEPA	1976–77	THMs, halogenated solvents, benzene, halobenzenes, halophenols, a haloether, PCBs and several PAHs	Finished water in 113 U.S. utilities; 4 sampling dates	Brass et al. 1977
National Statistical Assessment of Rural Water Conditions	Cornell Univ. (under contract with USEPA)	1978–79	Color, turbidity, TDS, hardness, Pb, Fe, Mn, Na, Mg on all; other metals, radionuclides, pesticides (endrin, lindane, methoxychlor, toxaphene, 2,4-D, 2,4,5-TP) on some	Individual & small community systems	Francis et al., 1984
Office of Radiation Programs Survey	USEPA	1980–81	8 radionuclides (^{226}Ra, ^{228}Ra, U ^{222}Rn, ^{210}Pb, ^{210}Po, ^{230}Th, ^{232}Th)	2500 U.S. public water systems; mostly larger GW systems	
TEAM Study	USEPA	1979–85	VOCs (mostly benzene, xylenes, styrene, ethylbenzene, C1 & C2 chlorohydrocarbons, p-dichlorobenzene) and pesticides	700 homes in 12 cities	Wallace 1987; Pellizzari et al. 1987a; Pellizzari et al. 1987b

Study	Organization	Date	Parameters	Coverage	Reference
National Inorganic and Radiologic Survey	USEPA	Early 1980s	Pb, Rn, U, and others	1000 community water systems	Not published
1984 Water Utility Operating Data	AWWA	1984	Turbidity, temperature, hardness, alkalinity and pH, all on raw water, last 3 on finished water also	430 of 600 largest U.S. utilities (serving >50,000)	AWWA 1986
1985 Water Utility Operating Data	AWWA	1985	Turbidity, temperature, hardness, alkalinity and pH, all on raw water, last 3 on finished water also	211 of next 600 largest U.S. utilities (serving 25,000–50,000)	AWWA 1987
Survey to Support Analysis of the Impacts of Proposed Regs. Concerning Filtration and Disinfection of Public Drinking Water Supplies	Association of State Drinking Water Administrators (through a grant from USEPA)	1986	Turbidity		
Water Industry Database	AWWA	1989–92	Temperature, pH, TDS, alkalinity, hardness, nitrate, Fe, Mn, color, turbidity, chorine and chloramine residuals and some TOC and TTHM	About 35% (1100) of U.S. utilities serving >10,000.	Water Industry Database 1992
National Pesticide Survey	USEPA	1988–90	101 pesticides, 25 pesticide degradates, and nitrate	1349 U.S. groundwater sources	USEPA 1990 USEPA 1992

From 1989 to 1992 the American Water Works Association (AWWA) assembled a "Water Industry Database" (WIDB) from questionnaires sent to all utilities serving more than 10,000 customers. Of the 600 utilities serving greater than 50,000 they received an 80% return on their survey, and a 26% return on the 2400 systems serving 10,000 to 50,000. They estimate that their returns represent about 50% of the people in systems serving greater than 10,000. Data collected include temperature, pH, total dissolved solids (TDS), alkalinity, hardness, nitrate, iron, manganese, color, turbidity, chorine and chloramine residuals and some information on total organic carbon (TOC) and total trihalomethanes (TTHM). A 120-page report dated May 1992 and entitled, "Water Quality Profiles: Groundwater, Surface Water, Distribution System" summarizes these data. A graphical summary and statistical analysis of the WIDB raw water quality data can be found in Letkiewicz et al. (1992). The AWWA resurveyed these same utilities in 1996, and the data are available in electronic format as WATER:\STATS from AWWA.

A variety of other national surveys of limited scope and sample size have been published. Among the more significant of these are the National Inorganic and Radiologic Survey, the National Pesticide Survey, the National Organics Reconnaissance Survey (NORS), and the National Organics Monitoring Survey (NOMS).

Although there are no national databases for drinking water quality, an Information Collection Rule (ICR), promulgated by the EPA Office of Water in 1996, will result in a limited national database. A wide range of water quality characteristics are expected to be collected from treated waters. All utilities serving 10,000 people or more must participate. Depending on the treatment processes used, data on chlorination byproducts, chloramination byproducts, ozonation byproducts, and chlorine dioxide byproducts will be collected and compiled. These will be available in a PC database format (MS Access).

3.4.2.2 Source-Specific Estimates

The determination of a source-specific average water quality or water quality distribution would naturally require some site-specific data. At the very least, one would need to know either the raw water quality and treatment conditions or the plant effluent quality. With the raw water quality and treatment conditions, one could estimate contaminant concentrations in the finished water at the entrance to the distribution system. Then, a distribution system model could be employed under average conditions for the network to predict some average water quality at the tap. Alternatively, a Monte Carlo simulation could be run to get a realistic distribution of qualities.

3.4.2.3 Site-Specific Estimates

Site-specific estimates can be made for contaminant concentrations at a single consumer's tap using distribution systems models (and possibly treatment models) as discussed above. Rather than using average distribution system characteristics, the specifics of the particular location would be used.

3.4.3   DATA GAPS AND RESEARCH NEEDS

The greatest weaknesses in estimating source water quality depend on the particular contaminant that is being examined. For the disinfection byproducts and corrosion byproducts, the greatest needs are in the area of distribution system modeling. Most other contaminants would benefit most from improved understanding of raw water concentrations. All contaminants are subject to variable removal by POU systems. Better knowledge of their actual performance would help in cases where POUs are important.

REFERENCES

AWWA (1986) 1984 Water Quality Operating Data. American Water Works Association, Denver, CO.

AWWA (1987) 1985 Water Quality Operating Data. American Water Works Association, Denver, CO.

Black, B. B., Harrington, G. W., and Singer, P. C. (1996) Reducing Cancer Risks by Improving Organic Carbon Removal. Journal American Water Works Association 88(6): 40–52.

Brass, H. J., Feige, M. A., Halloran, T. et al. (1977) The National Organic Monitoring Survey: Samplings and Analyses for Purgeable Organic Compounds. In Pojasek, R. B. (Editor). Drinking Water Quality Enhancement Through Source Protection. Ann Arbor Science Publ., Inc., Ann Arbor, MI, pp 393–416.

Francis, J. D., Brower, B. L., Graham, W. F., Larson, O. W., McCaull, J. L., and Vigorita, H. M. (1984) National Statistical Assessment of Rural Water Conditions. USEPA Office of Drinking Water, EPA 570/9-84-004, Washington, D.C.

Letkiewicz, F. J., Grubbs, W., Lustik, M., Cromwell, J., Mosher, J., Zhang, X., and Regli, S. (1992) Simulation of Raw Water and Treatment Parameters in Support of the Disinfection Byproducts Regulatory Impact Analysis. In Proceedings of the 1992 AWWA Annual Conference, American Water Works Association, Denver, CO, pp 1–48.

Pellizzari, E. D., Perritt, K., Hartwell, T. D. et al. (1987a) Total Exposure Assessment Methodology (TEAM) Study: Elizabeth and Bayonne, New Jersey: Devils Lake, North Dakota; and Greensboro, North Carolina. Vol. II. U.S. Environmental Protection Agency, Washington, D.C.

Pellizzari, E. D., Perritt, K., Hartwell, T. D. et al. (1987b) Total Exposure Assessment Methodology (TEAM) Study: Selected Communities in Northern and Southern California. Vol. III. U.S. Environmental Protection Agency, Washington, D.C.

Stumm, W. R. and Morgan, J. J. (1996) Aquatic Chemistry: Chemical Equilibria and Rates in Natural Waters, John Wiley & Sons, New York.

Symons, J. M., Bellar, T. A., Carswell, J. K. et al. (1975) National Organics Reconnaissance Survey for Halogenated Organics. Journal American Water Works Association 67: 634–647.

US EPA (1990) National Survey of Pesticides in Drinking Water Wells, Phase 1 Report. U.S. Environmental Protection Agency, Washington, D.C.: Report No. EPA 570/9-90-015.

US EPA (1992) Another Look: National Survey of Pesticides in Drinking Water Wells, Phase 2 Report. U.S. Environmental Protection Agency, Washington, D.C.: Report No. EPA 570/9-91-020.

Wallace, L. A. (1987) Total Exposure Assessment Methodology (TEAM) Study: Summary and Analysis, Vol. I. U.S. Environmental Protection Agency, Washington, D.C.

Water Industry Database (1992) Water Quality Profiles: Groundwater, Surface Water, Distribution System. American Water Works Association, Denver, CO Report No.: May 1992.

Weisel, C. P., and Chen, W. J. (1994) Exposure to Chlorination By-Products from Hot Water Uses. Risk Analysis 14:101–106.

WITAF (1993) Mathematical Modeling of the Formation of THMs and HAAs in Chlorinated Natural Waters, American Water Works Association, Denver, CO.

4 Developing Exposure Estimates

Clifford P. Weisel, John C. Little, Nancy Chiu, Spyros N. Pandis, Cliff Davidson, and Charles R. Wilkes

CONTENTS

1-57881-001-9/99/$0.00+$.50
© 1999 by International Life Sciences Institute

4.1 INTRODUCTION/OVERVIEW

Clifford P. Weisel

Exposure to contaminants in drinking water occurs not only via ingestion but also via inhalation, when the contaminants volatilize or are aerosolized, and via dermal absorption from direct contact with the water. Modeling the exposure and internal dose for waterborne contaminants within a home therefore needs to consider the properties of the contaminant, the three main exposure routes (ingestion, inhalation, and dermal absorption), release of the contaminant from water into the air within different microenvironments in the home, and the activities of individuals that result in exposures as they move through these different microenvironments. Releases of contaminants into the indoor air, which are subsequently transported to all rooms of the home occur during washing, showering, and bathing and the operation of toilets, dishwashers, humidifiers, and washing machines. Model parameters include the expected water concentration, activities engaged in by individuals, and the characteristics of the house that can influence the water and air concentrations. Exposure to waterborne contaminants can be quantified as distinct values or as a distribution of values for the community affected, since the model parameters indicated vary across a population.

4.1.1 INGESTION EXPOSURE

Although ingestion exposure was not considered in detail within this document, the following is provided as a basis for comparison with inhalation and dermal exposures. The ingestion exposure and dose are calculated based upon the water concentration, an estimate of the amount of water ingested, and the absorption efficiency across the gut. A total water ingestion of 2 L/day per person has been used by EPA in risk calculations. However, this value includes ingestion of all fluids. Some of the water ingested is heated in cooking or in the preparation of hot drinks, such as coffee and tea, and heating water can alter the concentration of the contaminants (for example, volatile compounds would be released from the water during boiling). The 2L/day value therefore can lead to an overestimation of the ingestion exposure to volatile contaminants in tap water, if contaminant losses are not taken into account. A mean ingestion of 1.4 L/day has been proposed for adults for tap water (U.S. EPA *Exposure Factors Handbook*) and a value of 1 L/day for children, who drink smaller quantities of liquids than adults and consume little water that has been heated for coffee or tea. The actual quantity of residential tap water consumed by an individual will vary according to that person's individual habits, the time spent at home, the season (temperature and humidity), and activities such as physical exertion. Thus, each situation must be considered independently, and regulation of a contaminant or advice given to the public when contamination of a water supply occurs should reflect the distribution of ingestion rates for the target population. The generic formula for calculating an internal dose for an ingestion exposure is:

$$Dose = C_w \times Amt \times AE$$

where C_w is the concentration of the water ingested (accounting for losses from volatilization and changes within the water during processing), Amt is the amount ingested, and AE is the absorption efficiency (fraction of the contaminant absorbed) across the GI tract.

4.1.2 INHALATION EXPOSURE

Inhalation exposure occurs when the air breathed contains compounds volatilized from water or aerosols formed during water usage such as bathing, showering, washing, cooking, etc. Showering has been identified as the activity contributing the greatest amount to inhalation exposure to volatile compounds (Wilkes et al. 1992), though some increase in the indoor air concentration of volatile water contaminants has been measured and modeled for a variety of water uses. Showering not only results in a direct inhalation exposure to the individual showering while in the shower stall and in the bathroom while the shower water is running, but it also increases the air concentration throughout the home, causing exposure to others within the residence. Other water uses, such as washing dishes or clothing, involve agitation of the water. These activities result in both volatilization of water contaminants and the production of aerosols. These processes have been studied more extensively for volatile water constituents than for aerosols containing nonvolatile species. Volatilization of a compound from water is a function of the amount of water used, the temperature of the water, and the volatility/solubility of the compound, usually expressed as the compound's Henry's Law constant. A number of models and experiments have been described in the literature that can be used to estimate the inhalation exposure and dose for volatile species associated with water use (McKone 1987, Wilkes et al. 1996, Georgopoulos et al. 1997).

The inhalation exposure for nonvolatile species, assuming that there is no partitioning of the species during the aerosol generation process, will be a function of the amount of aerosols produced, the water concentration, the size distribution of the aerosol, growth or shrinkage of the aerosols, and aerosol transport within the home. Impaction of the aerosols (>1 µm) onto surfaces is expected to be a more efficient removal process from the indoor air than it is for volatile species. To estimate inhalation exposures to nonvolatile species, the models developed for volatile compounds would have to be modified to account for aerosol generation at the water source (e.g., shower), formation of aerosols within a high humidity area (e.g., bathroom), changes in particle size distribution, and removal mechanisms during transport through the house. Modeling of inhalation exposures also should incorporate the activities of the individual being exposed (water use, shower/bathing duration and temperature, time spent indoors at home), water concentration, characteristics of the home (air exchange rates between rooms and outdoors, presence of vents in the bathroom, location of sinks, washing machines, etc.) and water use by other individuals within a home.

4.1.3 DERMAL ABSORPTION

Skin contact with contaminated water during showering, bathing, and swimming has been demonstrated to result in the penetration of the contaminants into the body (Jo et al. 1990, Weisel and Shephard 1994, Weisel and Jo 1996). Detailed modeling of how contaminants penetrate the skin has been attempted in order to predict dermal absorption (see Chapter 6). However, few actual measurements of the dermal dose following showering have been made. The penetration of the skin is a function of the permeability coefficient of a compound, which has often been estimated from the octanol/water partition coefficient. The primary activities that have been related to direct dermal absorption are bathing and showering, though other activities such as washing and handling wet clothing can be contributors. The duration of the activity and area of skin surface exposed during washing and handling wet clothing suggest that they will have a minor contribution to dermal absorption.

Few data are available on dermal absorption related to contact with vapors and aerosols. The dermal absorption of a contaminant from the vapors and aerosols within and outside of a shower stall is predicted to be much lower than that resulting from direct contact with the water.

REFERENCES

Bunge, A., Flynn, F. L., and Guy, R. H. (1994) Predictive Model for Dermal Exposure Assessment. In Wang, R. (Editor) *Water Contamination and Health,* Marcel Dekker, New York, pp 347.

Georgopoulos, P.G., Walia, A., Roy, A., and Lioy, P. J. (1997) Integrated Exposure and Dose Modeling and Analysis System. 1. Formulation and Testing of Microenvironmental and Pharmacokinetic Components. *Environmental Science & Technology* 31, 17–27.

Giardino, N, Gumerman, E., Andelman, J., Wilkes, C., and Small, M. (1990) Real-time Air Measurements of Trichloroethylene in Domestic Bathrooms Using Contaminated Water. In: Proc. 5th Int. Conf. Indoor Air Quality and Climate, Toronto, July 1990, 2, 707.

Jo, W., Weisel, C., and Lioy, P. (1990) Chloroform Exposure and the Health Risk Associated with Multiple Uses of Chlorinated Tap Water. *Risk Analysis* 10, 581–585.

Lindstrom, A., Highsmith, V., Buckley, T., Pate, W., and Michael, L. (1994) Gasoline-Contaminated Ground Water as a Source of Residential Benzene Exposure: A Case Study. *J. of Exposure Analysis and Environmental Epidemiology* 4, 183–185.

McKone, T. (1987) Human Exposure to Volatile Organic Compounds in Household Tap Water: The Indoor Inhalation Pathway. *Environmental Science & Technology* 21, 1194–1201.

Weisel, C.P. and Jo, W.-K. (1996) Ingestion, Inhalation and Dermal Exposure to Chloroform and Trichloroethene from Tap Water, *Environmental Health Perspectives* 104, 48–51.

Weisel, C., and Shepard, T. (1994) Chloroform Exposure and Body Burden Associated with Swimming in Chlorinated Pools. In Wang, R. (Editor) *Water Contamination and Health*, Marcel Dekker, New York, pp 135.

Wilkes, C., Small, M., Andelman, J., Giardino, N., and Marshall, J. (1992) Inhalation Exposure Model for Volatile Chemical from Indoor Uses of Water. *Atmospheric Environment* 26a, 2227.

Wilkes, C., Small, M., Davidson, C., and Andelman, J.(1996) Modeling the Effects of Water Usage and Co-Behavior on Inhalation Exposures to Contaminants Volatilized from Household Water. *J. of Exposure Analysis and Environmental Epidemiology* 6, 393.

4.2 TRANSFER OF VOLATILE COMPOUNDS FROM DRINKING WATER TO INDOOR AIR

John C. Little and Nancy Chiu

4.2.1 INTRODUCTION

The potential for the volatilization of contaminants from drinking water to indoor air was first identified by Prichard and Gesell (1981) in the context of radon. Those authors reported radon transfer efficiency (TE) as 0.90 for cleaning, dishwashing, and laundry, 0.63 for showers, 0.47 for baths, and 0.30 for toilets and kitchen use (Prichard and Gesell, 1981). A subsequent study by Hess et al. (1982) showed a strong correlation between concentrations of radon in air and in household water supplies, further verifying this exposure pathway. A detailed review and assessment of the potential for drinking water to serve as a source of airborne radon was provided by Nazaroff et al. (1987).

Attention was subsequently focused on the release of volatile organic compounds from drinking water (Andelman, 1985a, b). In the ensuing years Andelman and co-workers carried out several additional experimental and field studies further elucidating this exposure route, the results of which have been summarized in a detailed review (Andelman, 1990). A number of additional studies have also been conducted. Research carried out on the volatilization from showers included detailed experimental studies by Hodgson et al. (1988), Jo et al. (1990a, 1990b), McKone and Knezovich (1991), Tancrede et al. (1992), Giardino, Esmen and Andelman (1992), and Giardino and Andelman (1996), as well as a field study by Stern and Andrews (1989). The work of Jo et al. (1990a, b) also considered dermal sorption while the study of Giardino, Esmen, and Andelman (1992) focused on the role of the drop-size distribution generated by the shower head. Related theoretical assessments were made by McKone (1987), Little (1992), and Wilkes et al. (1992, 1996). Volatilization from showers has been studied most extensively, primarily because it has been shown to represent the most significant contributor of the various water use scenarios (McKone, 1987). More recently, however, experimental research has been reported on washing machines (Shepherd et al. 1994) and kitchen sinks (Howard and Corsi, 1995).

4.2.2 HENRY'S LAW AND THE MASS-TRANSFER COEFFICIENT

Phase equilibrium between air and a dilute aqueous solution of dissolved gas is given by Henry's Law as

$$y = HC \qquad (4.2.1)$$

where y is the gas-phase contaminant concentration, C the equilibrium aqueous-phase contaminant concentration, and H is a dimensionless Henry's Law constant. The temperature dependence of this dimensionless form of Henry's constant (Selleck et al., 1988) is

$$H \propto (1/T)10^{-(B/T)} \qquad (4.2.2)$$

where B is a temperature correction coefficient and T is the absolute temperature. Values for H and B for a wide range of volatile compounds are listed in Table 4.2.1.

Where air and water are brought into contact, the mass-transfer flux between the two phases, J, is proportional to the prevailing concentration driving force, or

$$J = K_{OL}(C - y/H) \qquad (4.2.3)$$

where K_{OL} is the overall mass-transfer coefficient based on liquid-phase concentrations, C is the bulk aqueous-phase concentration, and y is the bulk gas-phase concentration. The quantity, y/H, converts the gas-phase concentration into the equivalent (equilibrium) aqueous-phase concentration. The two-resistance theory (Lewis and Whitman, 1924) relates the overall mass-transfer coefficient to the individual liquid-phase and gas-phase mass-transfer coefficients, K_L and K_G, respectively:

$$\frac{1}{K_{OL}} = \frac{1}{K_L} + \frac{1}{HK_G} \qquad (4.2.4)$$

The mass-transfer flux, defined in Equation 4.2.3, must pass through the interface of area A existing between the water and the air. Since this active interfacial area is seldom easy to measure directly, it is usually lumped together with the mass-transfer coefficient, and experimental measurements are made to determine the combined parameter. The two-resistance theory still applies with Equation 4.2.4 modified as follows:

$$\frac{1}{K_{OL}A} = \frac{1}{K_L A} + \frac{1}{HK_G A} \qquad (4.2.5)$$

It is common to normalize the interfacial area by the active volume of water, V_L, from which mass transfer is occurring. When based on this specific interfacial area a, Equation 4.2.5 becomes

$$\frac{1}{K_{OL}a} = \frac{1}{K_L a} + \frac{1}{HK_G a} \qquad (4.2.6)$$

Whether to use Equation 4.2.4, 4.2.5, or 4.2.6 depends on whether it is possible to directly determine values of A and V_L by experiment, which in turn depends on the system being analyzed.

The values of K_L and K_G depend on the degree of turbulence, the fluid temperature, and the diffusion coefficient of the compound being transferred in each of the respective phases. Theoretical and experimental studies indicate that turbulent mass-transfer coefficients generally depend on the diffusion coefficient raised to some power, the magnitude of which lies between 1/2 and 2/3 (Little, 1992). If no better information is available, the following relationships are often assumed (Little, 1992):

$$K_L \propto D_L^{1/2} \qquad (4.2.7)$$

and

$$K_G \propto D_G^{2/3} \qquad (4.2.8)$$

where D_L and D_G are the diffusion coefficients of the compounds in water and air, respectively. These relationships provide a means to adjust a mass-transfer coefficient measured for one compound to the equivalent mass-transfer coefficient for another compound, provided the diffusion coefficients for the two compounds are known. Table 4.2.1 provides information from which D_L and D_G may be obtained. The gas-phase Schmidt number $Sc_G = (\mu_G/\rho_G D_G)$ is essentially independent of temperature and together with the density (ρ_G) and viscosity (μ_G) of air gives D_G at any temperature. D_L is obtained from the diffusion coefficient for oxygen in water at 20°C and adjusted to other temperatures using the viscosity of water (μ_L) as indicated.

4.2.3 PREDICTING MASS-TRANSFER RATES

Prediction of the rate of mass transfer between water and air requires an assumed flow model. This accounts for the variation in concentration driving force over the entire air/water interfacial area. In keeping with the concept of microenvironments, it will be assumed that the volume of air in contact with the water constitutes a completely mixed microenvronment. It therefore remains to select the most appropriate

TABLE 4.2.1
Volatile Compound Data (Selleck et al., 1988)

Compound (abbreviation)	H* (20°C)	B (K)	Sc_G (–)	$D_L/D_L(O_2)$** (–)
1,1,2-Trichlorotrifluoroethane	9.9	1600	2.3	0.39
Radon	3.9	1340	1.3	0.57
Fluorotrichloromethane	3.0	1030	1.8	0.40
Carbon tetrachloride	0.88	1820	2.1	0.41
1,1,1-Trichloroethane (TCA)	0.57	1770	1.9	0.40
Tetrachloroethylene (PCE)	0.54	1930	1.9	0.38
Trichloroethylene (TCE)	0.32	1960	1.8	0.42
Ethylbenzene	0.28	1700	2.0	0.36
trans-1,2-Dichloroethylene	0.30	1820	1.6	0.50
m- & p-Xylenes	0.22	1900	2.0	0.39
Toluene	0.22	1630	1.8	0.40
1,1-Dichloroethane	0.18	1800	1.6	0.47
o-Xylene	0.16	1900	2.0	0.39
cis-1,2-Dichloroethylene	0.12	1820	1.5	0.50
m-Dichlorobenzene	0.12	1600	2.1	0.36
Chloroform	0.12	1930	1.5	0.46
Monochlorobenzene	0.11	1500	1.9	0.40
1,2-Dichloropropane	0.092	1620	1.8	0.41
Methylene chloride	0.090	1620	1.4	0.57
Dichlorobromomethane	0.067	2050	1.6	0.41
o-Dichlorobenzene	0.047	1600	2.1	0.36
1,2-Dichloroethane	0.040	1540	1.7	0.47
Ethylene dibromide	0.040	2000	1.7	0.42
Chlorodibromomethane	0.036	2050	1.8	0.43
1,1,2-Trichloroethane	0.028	1600	1.8	0.40
Bromoform	0.017	2170	2.0	0.41
1,1,2,2-Tetrachloroethane	0.012	1600	2.4	0.39
1,2-Dibromo-3-chloropropane (DBCP)	0.0056	2350	2.7	0.34
Ammonia	0.00055	1890	0.64	0.96

* $H \propto (1/T)10^{-(B/T)}$

** $D_L(O_2) = 2.24 \times 10^{-9}$ m²/s at 20°C and $D_L \propto (T/\mu_L)$

flow model to represent the water in contact with the air in the microenvironment. Clearly, the type of flow model selected will depend on the nature of the water using device. For example, volatile emissions from a shower have been analyzed as a plug flow stream of water, whereas emissions from a bath might be evaluated assuming that the volume of water is well mixed. The process of combining the expression for the mass-transfer flux with an appropriate flow model will be illustrated by describing two idealized cases, one for mass transfer from a plug flow stream of water and one for mass transfer from a completely mixed volume of water.

Consider the hypothetical situation shown schematically in Figure 4.2.1. The volumetric flow rates of water (Q_L) and air (Q_G) are assumed to be constant with

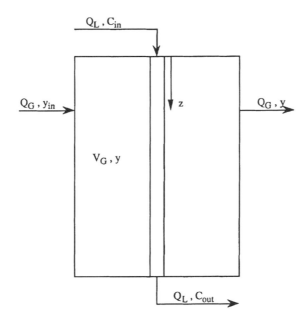

FIGURE 4.2.1 Schematic representation of a plug flow stream of water in contact with a completely mixed volume of air.

time and the air volume (V_G) is assumed to be well mixed. The air entering the microenvironment has constant contaminant concentration y_{in} (usually $y_{in} = 0$), and at time $t = 0$ the flow of water is started. The concentration of the contaminant entering in the water is C_{in}. As the water passes through the air, it loses contaminant at a rate proportional to the concentration driving force existing between the water and the air as given by Equation 4.2.3. Assuming the water is in plug flow, a differential mass balance on the contaminant in the water passing through the system taken in the z-direction gives

$$\frac{dC}{dz} = -\frac{JP}{Q_L} = -\frac{K_{OL}(C - y/H)P}{Q_L} \tag{4.2.9}$$

where P is the perimeter of the water stream, C is the concentration in the water, and y is the concentration in the air. If the stream of water passes through the compartment rapidly in relation to the time for the gas-phase concentration to change, we can assume that y is effectively constant during the residence time of the water. In general, this condition is referred to as the pseudo-steady-state assumption. Equation 4.2.9 may then be integrated to obtain

$$C_{out} = C_{in} \exp\left(-\frac{K_{OL}A}{Q_L}\right) + \left(\frac{y}{H}\right)\left(1 - \exp\left(-\frac{K_{OL}A}{Q_L}\right)\right) \tag{4.2.10}$$

where $A = PL$ is the active interfacial area through which the mass-transfer flux passes, and L is the length of the stream of water. A transient mass-balance on the contaminant in the gas-phase microenvironment results in

$$\frac{dy}{dt}V_G = Q_L\left(C_{in} - C_{out}\right) - Q_G\left(y - y_{in}\right)$$

(4.2.11)

Substituting Equation 4.2.10 into Equation 4.2.11 and integrating gives

$$y = \frac{s}{r} + \left(y_o - \frac{s}{r}\right)\exp(-rt)$$

(4.2.12)

where

$$s = \frac{\left(Q_L C_{in}\left(1 - \exp\left(-\frac{K_{OL}A}{Q_L}\right)\right) + Q_G y_{in}\right)}{V_G}$$

and

$$r = \frac{\left(\left(\frac{Q_L}{H}\right)\left(1 - \exp\left(-\frac{K_{OL}A}{Q_L}\right)\right) + Q_G\right)}{V_G}$$

Equations 4.2.10 and 4.2.12 simplify under certain limiting conditions. When $t \to \infty$, $y \to s/r$, which is the steady-state concentration. Also, when $y = 0$, $C_{out} = C_{in}\exp(-K_{OL}A/Q_L)$.

Contrast the previous situation with that shown schematically in Figure 4.2.2, where the volume of water in contact with the air is now considered to be perfectly mixed. A transient mass-balance on the contaminant in the water yields

$$\frac{dC}{dt}V_L = Q_L C_{in} - Q_L C - JA$$

(4.2.13)

while a similar balance for the air gives

$$\frac{dy}{dt}V_G = Q_G y_{in} - Q_G y + JA$$

(4.2.14)

Substituting Equation 4.2.3 for the mass transfer flux gives

$$\frac{dC}{dt}V_L = Q_L C_{in} - Q_L C - K_{OL}A(C - y/H)$$

(4.2.15)

and

$$\frac{dy}{dt}V_G = Q_G y_{in} - Q_G y + K_{OL}A(C - y/H)$$

(4.2.16)

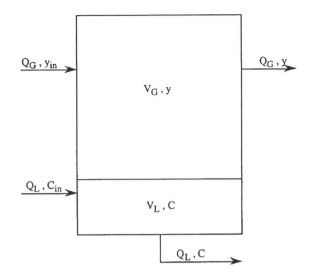

FIGURE 4.2.2 Schematic representation of a completely mixed volume of water in contact with a completely mixed volume of air.

Thus both the plug flow model (Equations 4.2.10 and 4.2.12) and the completely mixed flow model (Equations 4.2.15 and 4.2.16) may be used to predict gas-phase and aqueous-phase concentrations as a function of time, providing reliable estimates for $K_{OL}A$, and the other required parameters are available. The two sets of equations are also used to determine mass-transfer coefficients from data collected in experimental studies.

4.2.4 MEASURING MASS-TRANSFER COEFFICIENTS

For plug flow systems the overall mass-transfer coefficient may be determined by measuring both C_{out} and y over a period of time and then determining $K_{OL}A$ by means of a least squares fit of Equations 4.2.10 and 4.2.12 to both data sets. Equations 4.2.15 and 4.2.16, when solved simultaneously, may be used in a similar fashion to determine values of $K_{OL}A$ from experimental data in completely mixed systems.

The two-resistance theory illustrates the relationship between the individual liquid- and gas-phase mass-transfer coefficients. For example, Equation 4.2.4 shows that the overall resistance to mass transfer ($1/K_{OL}$) equals the sum of two resistances in series: one for the liquid phase ($1/K_L$) and one for the gas phase ($1/HK_G$). For highly volatile compounds (large H) the gas-phase resistance is negligible and the rate of mass transfer is controlled by the liquid-phase resistance. In contrast, for compounds of low volatility (small H), the gas-phase resistance controls. To predict the rate of mass transfer for a series of compounds spanning a wide range of volatility, reliable estimates are required of both the liquid-phase and gas-phase mass-transfer coefficients.

Earlier workers have found that the K_G/K_L ratio appears to be reasonably constant under similar conditions for a given mass-transfer system (for example, see Munz and Roberts, 1989). If the ratio is known, it can be used to obtain an estimate of either K_LA or K_GA when only the other is known. Having an estimate of the K_G/K_L ratio is especially valuable as, in many cases, $K_{OL}A$ is measured using only highly volatile compounds. Since liquid-phase resistance usually dominates under these conditions, the value of $K_{OL}A$ inferred from the experimental data is essentially equal to the value of K_LA. To predict the extent of volatilization of low volatility compounds, an estimate of K_GA is required. The only way to obtain this is either through additional experimental measurements, or by calculation using an estimate of the K_G/K_L ratio. The latter approach is clearly less accurate, but may suffice for a first approximation.

Mass transfer coefficients typically vary with liquid and gas flow rates, temperature, and the diffusion coefficients of the transferring chemicals within each of the fluids. If two or more chemicals of differing volatility are transferred under identical hydrodynamic and temperature conditions and through the same interfacial area, then all of these variables remain constant except the diffusivities. This provides a basis for evaluating the individual liquid- and gas-phase mass-transfer coefficients from experimentally determined values of the overall mass-transfer coefficient, $K_{OL}A$ (Cho and Wakao, 1988). Providing a series of compounds are selected that have similar values for D_L and D_G, the individual mass-transfer coefficients K_LA and K_GA can be evaluated from the intercept and slope of a plot of $1/K_{OL}A$ vs. $1/H$, as shown by Equation 4.2.5.

4.2.5 Relationship between $K_{OL}A$, $K_{OL}a$, and TE

A similar development of the plug flow and completely mixed flow models could have led to the following alternate expression for the terms involving the mass-transfer coefficient in Equations 4.2.10, 4.2.12, 4.2.15, and 4.2.16:

$$\frac{K_{OL}A}{Q_L} = \frac{K_{OL}A\theta}{V_L} = K_{OL}a \cdot \theta \qquad (4.2.17)$$

where θ is the residence time of the water in the system and V_L is the active volume of water within the system. Equation 4.2.17 provides a relationship for converting between the two forms of the mass-transfer coefficient, $K_{OL}A$ and $K_{OL}a$.

In the past, most information on volatilization from drinking water to indoor air has been presented in the form of a transfer efficiency (for example, Prichard and Gesell, 1981). This approach is limited in that TE only applies to a specific volatile compound. To use the proposed approach involving mass-transfer coefficients, published values for TE can be translated into equivalent values for the mass-transfer coefficient. This approach will be useful until such time as mass-transfer coefficients have been measured for the various water using devices over a range of operating conditions. Transfer efficiency may be defined as the fractional volatilization when the gas-phase concentration is zero and the system is at steady-state. Therefore,

$$\frac{C_{out}}{C_{in}} = (1 - TE) \qquad (4.2.18)$$

Under steady-state conditions and with the gas-phase concentration equal to zero, the plug flow model yields

$$\frac{C_{out}}{C_{in}} = \exp\left(-\frac{K_{OL}A}{Q_L}\right) \qquad (4.2.19)$$

or

$$K_{OL}A = -Q_L \ln(1 - TE) \qquad (4.2.20)$$

The completely mixed flow model at steady-state and with $y = 0$ gives

$$\frac{C_{out}}{C_{in}} = \frac{1}{\left(1 + \dfrac{K_{OL}A}{Q_L}\right)} \qquad (4.2.21)$$

or

$$K_{OL}A = Q_L\left(\frac{TE}{1 - TE}\right) \qquad (4.2.22)$$

4.2.6 NONIDEAL CONDITIONS

Real water-using devices will clearly deviate from the assumed ideality of the plug flow and completely mixed flow models. This deviation can be handled in one of two ways. Either a more complicated nonideal flow model is developed by means of a tracer study, or the system is analyzed as though it is ideal with the nonideality being incorporated into the values of $K_{OL}A$ estimated from the experimental data.

In certain circumstances, there may be several mass-transfer mechanisms occurring simultaneously within a single system. For example, in a shower where drops are formed, there are at least four distinct regions of mass transfer: drop formation, a period of drop acceleration to terminal velocity, the fall of the drop at terminal velocity through the shower, and coalescence on impact at the base of the shower. Clearly, the mass-transfer coefficient inferred from such experimental data will lump the contributions of the different regions into a single average parameter. Theoretically this assumption is acceptable, since the volatilization process is linear and, where a number of independent linear processes are occurring simultaneously, the overall effect is also linear (Levenspiel, 1972).

4.2.7 CURRENT STATE OF KNOWLEDGE

Much of the experimental work completed thus far has focused on the measurement of transfer efficiency, and not on the determination of individual gas- and liquid-phase mass-transfer coefficients. However, in three recent studies, mass-transfer

TABLE 4.2.2
Individual Mass-Transfer Coefficients for Different Shower Systems (Little, 1992)

Shower System	Water Flow Rate (L/min)	Temperature (°C)	$K_{OL}A$ (L/min)	K_LA (L/min)	K_GA (L/min)	K_G/K_L
1	5	42–46	8.9–1.9	9.5	130	13
2	8.7	40	2.6–2.4	2.5		
3	9.5	22	8.6	8.6		
3	9.5	37	9.7	9.7		
4	13.4	42	21.0–6.0	17	380	22
4	13.5	33	20.9–4.3	18	320	17
5	13.7	40	34–21	28		

TABLE 4.2.3
Individual Mass-Transfer Coefficients for a Kitchen Sink Operated under Different Conditions (Derived from the data of Howard and Corsi, 1995)

Conditions	Water Flow Rate (L/min)	Temperature (°C)	$K_{OL}a$ (1/min)	K_La (1/min)	K_Ga (1/min)	K_G/K_L
No aerator	4.8	25	.023–.0047	0.022	4.0	180
No aerator	7.9	25	.030–.0034	0.023	2.6	110
Screen	4.8	25	.018–.0016	0.016	1.2	75
Screen	7.9	25	.028–.0018	0.026	1.3	50
Aerator	4.8	25	.032–.0013	0.028	0.91	33
Aerator	6.3	25	.043–.0018	0.035	1.3	37
Aerator	7.9	25	.069–.0026	0.051	1.8	35
Dishes	4.8	25	.037–.0012	0.030	0.83	28
Dishes	6.3	25	.058–.0042	0.042	3.1	74

coefficients for showers (Little, 1992), a washing machine (Shepherd et al., 1994), and kitchen sinks (Howard and Corsi, 1995), have been evaluated. The results of the shower and kitchen sink studies are summarized in Tables 4.2.2 and 4.2.3, respectively. A value of $K_{OL}A = 0.98$ L/min was measured by Shepherd et al. (1994) for mass transfer of chloroform from a washing machine.

As shown in Table 4.2.2, different shower systems appear to have quite different mass-transfer coefficients, most probably influenced by the design of the shower head or nozzle as well as the different water flow rates. For systems 1 and 4 a fairly wide range of $K_{OL}A$ values are shown. The variation is a result of the wide range of volatile compounds studied in the experiments and illustrates the importance of gas-phase resistance in showers. These data were used to calculate individual liquid- and gas-phase mass-transfer coefficients, as indicated. Typical values for K_LA, K_GA, and K_G/K_L in showers are 15 L/min, 250 L/min, and 17, respectively.

TABLE 4.2.4

Preliminary (order-of-magnitude) estimates of K_LA and K_GA for Various Water-Using Devices under Typical Operating Conditions

System	Assumed Flow Model	Estimation Method	Water Flow Rate (L/min)	K_G/K_L	K_LA (L/min)	K_GA (L/min)
Shower	PFM	Literature	10	17	15	250
Faucet	PFM	Literature	5	63	1.5	100
Wash mach	CMFM	Literature	—	40*	1	40
Dishwasher	CMFM	TE = 0.90	1	40*	9	360
Bath	CMFM	TE = 0.47	8	40*	7	280
Toilet	CMFM	TE = 0.30	0.2	40*	0.07	2.8

* Assumed value

Table 4.2.3 shows mass-transfer coefficients determined for water flowing from a faucet in a kitchen sink for a range of water flow rates and operating conditions. Again, the wide range in $K_{OL}a$ values is due to the wide range of volatile compounds studied, allowing the calculation of individual mass-transfer coefficients as indicated. The observed values of K_La or K_Ga are converted to equivalent K_LA or K_GA values using the water flow rate, a system volume of 50 L, and Equation 4.2.17. Thus typical values for K_LA, K_GA, and K_G/K_L for a faucet/sink are 1.5 L/min, 100 L/min, and 63, respectively. This range in values shows that mass-transfer coefficients for faucets are somewhat lower than those measured in showers, while those for washing machines are about one order-of-magnitude lower than for showers. In general, the results indicate that mass-transfer coefficients vary substantially for different water-using devices as well as for different operating conditions within each device.

To apply the models for estimating the extent of volatilization from household water, estimates of individual mass-transfer coefficients are required for each water-using device at the specified operating conditions. Since no published information is available for mass-transfer coefficients in dishwashers, baths, and toilets, an estimate was obtained using the reported transfer efficiencies (TEs) for radon (Prichard and Gesell, 1981). In addition, an appropriate flow model had to be selected in order to calculate $K_{OL}A$ from TE using Equation 4.2.20 or 4.2.22. Once $K_{OL}A$ was determined, it was assumed equal to K_LA (since radon is highly volatile) and an estimate of K_GA was obtained using a K_G/K_L ratio of 40 (assumed to be equal to the average value for showers and faucets). The estimated values of the mass-transfer coefficients for each of the water using devices are summarized in Table 4.2.4. These estimates are preliminary and should only be considered accurate to within an order of magnitude.

4.2.8 FUTURE WORK

Future work should concentrate on measuring mass-transfer coefficients and on determining appropriate flow models for the various indoor water-using devices. Measuring mass-transfer coefficients over a wide range of contaminant volatility will enable the measured overall coefficient to be separated into component liquid- and gas-side coefficients. Determining these values over a wide range of operating conditions (for example, different types of shower head, changing water flow rate, and a range of water temperatures) will allow the natural variability of the parameters to be evaluated. The development of probability distributions that account for both the uncertainty of the measured values as well as the inherent natural variability is becoming more common when estimating exposure (EPA, 1995). Subsequently, an iterative process of exposure modeling and sensitivity analysis can be used to identify those variables that have the greatest influence on exposure and to determine whether additional research is required to improve the exposure model. For example, the magnitude of the Henry's Law constant will have a large impact on the extent of the inhalation exposure. Fortunately, methods of measuring Henry's Law constant are well established, except for compounds of extremely low volatility, and reliable values for most volatile compounds found in drinking water are usually available. Thus, further research could most profitably focus on the measurement of mass-transfer coefficients for those indoor water using devices that are identified as contributing most substantially to inhalation exposure.

REFERENCES

Andelman, J. B., Inhalation exposure in the home to volatile organic contaminants of drinking water, *The Science of the Total Environment*, 47, 443–460, 1985a.

Andelman, J. B., Human exposures to volatile halogenated chemicals in indoor and outdoor air, *Environmental Health Perspectives*, 62, 313–318, 1985b.

Andelman, J. B., Total exposure to volatile organic compounds in potable water, in *Significance and treatment of volatile organic compounds in water supplies*, Ram, N., Christman, R., and Cantor, K., Eds., Lewis Publishers, Chelsea, MI, 485–504, 1990.

Cho, J. S. and Wakao, N. J., Determination of liquid-side and gas-side volumetric mass-transfer coefficients in a bubble column, *Journal of Chemical Engineering Japan*, 21, 576–581, 1988.

EPA, Uncertainty analysis of risks associated with exposure to radon in drinking water, EPA Office of Water, EPA Report No. 822-R-96-005, 1995.

Giardino, N. J. and Andelman, J. B., Characterization of the emissions of trichloroethylene, chloroform and 1,2-dibromo-3-chloropropane in a full-size, experimental shower, *Journal of Exposure Analysis and Environmental Epidemiology*, 6, 413–423, 1996.

Giardino, N. J., Esmen, N. A., and Andelman, J. B., Modeling volatilization of trichloroethylene from a domestic shower spray: the role of drop size distribution, *Environmental Science & Technology*, 26, 1602–1606, 1992.

Hess, C. T., Weiffenbach, C. V., and Norton, S. A., Variations of airborne and waterborne Rn-222 in houses in Maine, *Environment International*, 8, 59–66, 1982.

Hodgson, A. T., Garbesi, K., Sextro, R. G., and Daisey, J. M., Lawrence Berkeley Laboratory Report No. LBL-25465, Berkeley, California, 1988.

Howard, C. and Corsi, R. L., Volatilization of chemicals from drinking water to indoor air: role of the kitchen sink, presented in EPA/AWMA's Engineering solutions to indoor air quality problems, Raleigh, North Carolina, 1995.

Jo, W. K., Weisel, C. P., and Lioy, P. J., Routes of chloroform exposure and body burden from showering with chlorinated tap water, *Risk Analysis*, 10, 575–580, 1990a.

Jo, W. K., Weisel, C. P., and Lioy, P. J., Chloroform exposure and the health risk associated with multiple uses of chlorinated tap water, *Risk Analysis*, 10, 581–585, 1990b.

Levenspiel, O., *Chemical Reaction Engineering*, 2nd ed., John Wiley & Sons, New York, 1972.

Lewis, W. K. and Whitman, W. G., Principles of gas absorption, *Industrial and Engineering Chemistry*, 16, 1215–1220, 1924.

Little, J. C., Applying the two-resistance theory to contaminant volatilization in showers, *Environmental Science & Technology*, 26, 1341–1349, 1992.

McKone, T. E., Human exposure to volatile organic compounds in household tap water: the indoor inhalation pathway, *Environmental Science & Technology*, 21, 1194–1201, 1987.

McKone, T. E. and Knezovich, J. P., The transfer of trichloroethylene (TCE) from a shower to indoor air: experimental measurements and their implications, *Journal of the Air & Waste Management Association*, 40, 282–286, 1991.

Munz, C. and Roberts, P. V., Gas- and liquid-phase mass transfer resistances of organic compounds during mechanical surface aeration, *Water Research*, 23, 589–601, 1989.

Nazaroff, W. W., Doyle, S. M., Nero, A. V., and Sextro, R. G., Potable water as a source of airborne Rn-222 in U S dwellings: a review and assessment, *Health Physics*, 52, 281–295, 1987.

Prichard, H. M. and Gesell, T. F., An estimate of population exposures due to radon in public water supplies in the area of Houston Texas, *Health Physics*, 41, 599–606, 1981.

Selleck, R. E., Mariñas, B. J., and Diyamandoglu, V., Sanitary Engineering and Environmental Health Research Laboratory, UCB/SEEHRL Report No. 88-3/1, University of California, Berkeley, California, 1988.

Shepherd, J., Kemp, J., and Corsi, R. L., Residential washing machines as sources of indoor air pollution — chloroform formation, mass transfer, and emission dynamics, presented in the proceedings of AWMA Annual Meeting, Cincinnati, Ohio, 1994.

Stern, A. H. and Andrews, L. R., The contribution of domestic water use to indoor air concentrations of chloroform in New York City apartments — a pilot study, *Toxicological and Environmental Chemistry*, 24, 71–81, 1989.

Tancrede, M., Yanagisawa, Y., and Wilson, R., Volatilization of volatile organic compounds from showers I. Analytical method and quantitative assessment, *Atmospheric Environment*, 26A, 1103–1111, 1992.

Wilkes, C. R., Small, M. J., Andelman, J. B., Giardino, N. J., and Marshall, J., Inhalation exposure model for volatile chemicals from indoor uses of water, *Atmospheric Environment*, 26A, 2227–2236, 1992.

Wilkes, C. R., Small, M. J., Davidson, C. I., and Andelman, J. B., Modeling the effects of water usage and co-behavior on inhalation exposures to contaminants volatilized from household water, *Journal of Exposure Analysis and Environmental Epidemiology*, 6, 393–412, 1996.

4.3   AEROSOLS AND WATER DROPLETS

Spyros N. Pandis and Cliff Davidson

4.3.1   DEFINITIONS (AEROSOLS, DROPLETS)

"Aerosols" are defined as small particles suspended in air or other gas (Friedlander, 1977). For the purposes of this study, we will define aerosol as an airborne particle

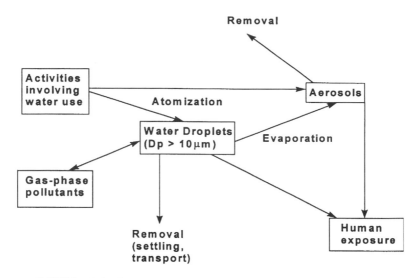

FIGURE 4.3.1   Schematic of aerosol/droplet processes and interactions.

sufficiently small (diameter $D_p \leq 10 \ \mu m$) that it does not rapidly settle out of the air. In this context we will refer to two groups of airborne particles, namely water droplets ($D_p > 10 \ \mu m$) and aerosols ($D_p \leq 10 \ \mu m$). Using the above definition, aerosols and water droplets have different physicochemical characteristics (solute concentrations, lifetimes, dynamics, etc.). Most uses of drinking water lead to the production of water droplets. These are relative large particles with small solute concentrations consisting mainly of water. After their production they are not in equilibrium with their environment, so rapid evaporation is expected. Due to their size, their lifetime is relatively short and if they do not evaporate rapidly they will be removed in a few minutes. During droplet evaporation their liquid water content decreases and solute concentrations increase. Aerosols produced by droplet evaporation are expected to be liquid for relative humidities exceeding 60% and ambient temperatures (Seinfeld and Pandis, 1997). During water evaporation the mass concentration of nonvolatile aerosol components (expressed in a per volume of air basis) remains constant. On the contrary, volatile species dissolved in tap water are transferred to the gas phase during droplet evaporation. This transfer is expected to be complete when all the water evaporates and the aerosol is dried. An overview of the processes related to the production of droplets, their evaporation to aerosols, and their removal is depicted in Figure 4.3.1. Droplets are too large to be respirable, so for inhalational exposure the aerosol concentration should be estimated. The droplets are important only as precursors of aerosols in this context. Aerosols and droplets may also contribute to dermal exposure if they contact the human skin.

Assuming that on average tap water contains 300 mg L^{-1} of dissolved solids (Owen et al., 1992) of 2.5 g cm^{-3} density, the diameter of the aerosol particle, d_a resulting in the complete evaporation of a droplet of diameter d_{dr} will be given by $d_a = 0.05 \ d_{dr}$. Therefore, for the purposes of estimating inhalational exposure to aerosols, droplets with diameters larger than 200 µm can be ignored. Larger droplets will produce nonrespirable ($d_a > 10$ µm) dry particles even if they completely evaporate. Note that droplets may shatter after impact with a solid or liquid surface (human body in a shower, sink, etc.) and smaller droplets may be created.

4.3.2   Water Uses as Sources of Droplets/Aerosols

Activities that possibly lead to production of droplets include (a) drinking water and beverage/foods prepared from tap water, (b) showering, (c) bathing, (d) personal washing (hands, face), (e) dish washing, (f) clothes washing, (g) toilet use, (h) humidifier use, (i) household cleaning, (k) cooking, and (l) outdoor water use.

Estimating the inhalational exposure to aerosols requires addressing the following questions:

1. What are the rates of droplet/aerosol production (mass of droplet per unit of time per mass of water used) for each activity? Do these rates depend on any other activity-related parameters (e.g., flux of water, showerhead type, etc.?) or can they be assumed to be constant?
2. What is the size distribution of these droplets?

Droplet and aerosol production in most of the above activities has been neglected in previous studies (see Owen et al., 1992 for a review) and there are no available data. Our goal here is to discuss the major processes leading to inhalational aerosol exposure and summarize the little available information about showering and use of humidifiers.

4.3.2.1   Showers

Exposure to chemical contaminants in a shower can occur from inhalation of aerosols, inhalation of gases, dermal contact, and swallowing of shower water. Although this section deals with the role of aerosols, it is important to note that inhalation of gases as well as inhalation of aerosols may be a strong function of the droplet size distribution, and that exposure through dermal contact and swallowing may also to some extent depend on particle size. In this section we consider droplet size distributions reported in the literature as a background for later sections on settling and evaporation.

The data available in the literature suggest that the droplet size distribution depends on the design of the shower head, the way the shower head is used (e.g., if there is an adjustable setting), and the water flow rate. One of the few studies to report shower droplet size distributions is that of Giardino et al. (1992). These investigators measured droplet sizes at water flow rates of 5 and 10 L min^{-1} for a single shower head. They showed that the distributions are bimodal at both flow rates. The droplets measured were larger than 200 µm in diameter. Note that the measured droplet distribution is the distribution produced by the shower head. Smaller droplets can be produced during the impaction of this spray with the body of the person in the shower.

4.3.2.2 Humidifiers

Highsmith et al. (1988, 1992) investigated the indoor aerosol concentrations associated with use of tap water in portable humidifiers. The size of the droplets generated
by an ultrasonic humidifier can be calculated a priori by application of the capillary
wave theory, in which capillary waves are established on a liquid surface in an
ultrasonic field by

$$d_d = 0.34 \left(\frac{8\pi\sigma_s}{\rho_l f^2} \right)^{1/3}$$

where d_d is the number mean diameter, σ_s is the liquid surface tension (72.7 dyn
cm^{-1} for water at 20°C), ρ_l is the liquid density in g cm^{-3}, and f is the ultrasonic
transducer frequency. This equation predicts a 3 μm number mean diameter (mean
diameter of an aerosol population weighted by particle number, see for example
Seinfeld (1986)) for water in a 1.65 MHz ultrasonic field, the frequency used by
many portable ultrasonic humidifiers (Highsmith et al., 1992). The mean size of the
solid particles created by evaporation of these droplets should theoretically be (using
a density of 2.5 g cm^{-3}) around 0.15 μm. The experimental data of Highsmith et al.
(1992) suggested lognormal distributions with number mean diameters from 0.19
to 0.34 and geometric standard deviations from 1.56 to 1.97.

4.3.3 Lifetime of Droplets and Aerosols

4.3.3.1 Removal

Droplets are removed mainly through gravitational settling and their "settling"
lifetime can be readily calculated. In this section we assume that the diameter of
the droplets remains constant during their settling. This diameter change due to
evaporation and its effect on the droplet lifetime will be discussed in the following
section. Equations describing the settling of spheres such as water droplets in a fluid
such as air have been well established. The settling velocity of unit density droplets
larger than 10 μm (Hinds, 1982) is depicted in Figure 4.3.2 and their lifetime (defined
as the ratio of the height where the droplets are produced over their deposition
velocity) is shown in Figure 4.3.3 for three heights. Droplets larger than 100 μm
settle in less than 10 s, while droplets smaller than 10 μm can survive for several
minutes.

     In cases where droplets evaporate to dryness, the resulting particle may be of
submicrometer size; a 1 μm diameter particle settles at 0.003 cm sec^{-1}, and thus may
remain airborne for a substantial period of time. Aerosols are removed by settling,
but also through diffusion and other processes (e.g., thermophoresis). Nazaroff et al.
(1989, 1990a, 1990b, 1991) have studied in detail these processes. Aerosol lifetimes
in the indoor environment are of the order of hours.

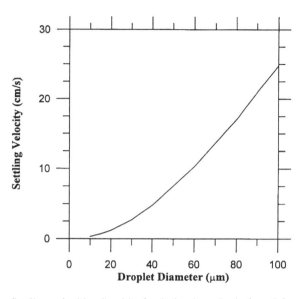

FIGURE 4.3.2   Settling velocities (in air) of unit density spherical particles as a function of their size.

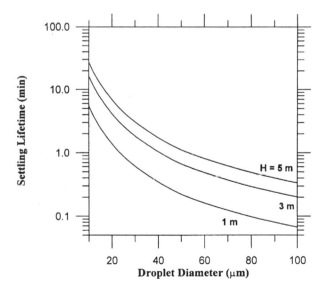

FIGURE 4.3.3   Lifetime of a water droplet due to gravitational settling as a function of its size and initial distance from the ground.

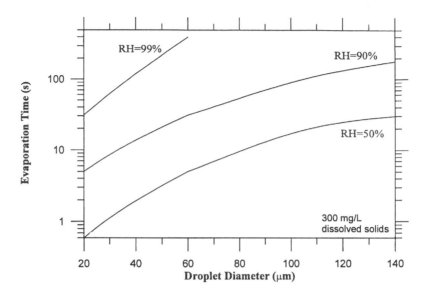

FIGURE 4.3.4   Estimated evaporation time (defined here as the time necessary for a droplet to reach a size less than 10 μm) as a function of initial droplet size and ambient relative humidity.

4.3.3.2   Evaporation

The time required for a water droplet to evaporate and become an aerosol particle (diameter smaller than 10 μm) is defined here as evaporation time. This time depends dramatically on the initial droplet diameter (smaller droplets evaporate faster) and on the ambient relative humidity. The evaporation times for droplets with a concentration of dissolved solids of 300 mg/L⁻¹ are depicted in Figure 4.3.4. These lifetimes have been calculated by solving numerically the water droplet growth/evaporation equation (Pruppacher and Klett, 1980). For relative humidities above 90% typical of the shower environments, this time scale is more than 1 minute for droplets larger than 80 μm. As the RH approaches 100% this time increases exponentially. For RH around 60% or lower even droplets as large as 150 μm can evaporate in less than 30 s.

Note that the resulting particles for relative humidities roughly above 50 to 60% are going to be liquid; that is, they will contain a small amount of water. Humidities lower than 50% are necessary for complete drying of these particles.

4.3.3.3   Evaporation and Removal

Because the settling velocity increases as droplet diameter increases, larger droplets will be airborne for a short time and will not produce aerosols, but they will rather be removed. This behavior implies that droplets larger than 50 μm can be safely neglected for aerosol exposure estimates in a shower or that droplets larger than 100 μm can be neglected in most indoor environments. As the few available studies have concentrated on measuring size distributions of droplets larger than 200 μm, this finding underlines the research need of measuring the production rate of smaller droplets that can produce aerosols. The effect of bounceoff from people's bodies in the shower, which decreases droplet diameter due to shattering of the droplets, is thus likely to be significant and probably cannot be neglected in estimating exposure (Gunderson and Witham, 1988).

4.3.4 INTEGRATION OF PROCESSES

4.3.4.1 Aerosol Production in Showers

Very little work has been done specifically to examine evaporation of droplets from showers. In one such set of experiments, Gunderson and Witham (1988) introduced a nonvolatile chemical species, ammonium fluorescein, in deionized shower water in concentrations varying from 5 to 50 ppm. They used two shower heads with heated water (104°F). Each shower head was operated at coarse- and fine-spray settings in order to obtain a range of conditions likely to be encountered in real showers. The water flow rate for one of the shower heads was specified by the manufacturer to be approximately 8 L min^{-1}, while the output of the other was not specified. A mannequin was positioned in the shower to allow bounceoff and sub-sequent shattering of the water droplets, similar to what would be encountered in a real shower. Airborne concentrations of respirable particles ($D_p < 10 \mu m$) were measured in the breathing zone. The measured concentrations represented the sum of both respirable water droplets and dried ammonium fluorescein particles. These results indicated that a nonvolatile species in water at 1 ppm would result in an airborne concentration of 0.03 to 0.2 μg m^{-3} in the breathing zone of the shower. The resulting aerosols in their experiments were mostly of submicrometer size, based on the concentrations introduced in their shower water, and thus were readily inhalable. It is also noteworthy that Gunderson and Witham (1988) reported the highest airborne particle concentrations for the highest water pressure and finest spray settings. They further commented that bounceoff from the surface of the mannequin most likely resulted in shattering of the droplets, implying production of respirable droplets from larger ones and enhancing evaporation to produce dry submicrometer aerosol. This study suggested that the expected inhalable particulate concentration for a *nonvolatile* species, c_a, at a concentration c_w can be approximated by:

$$c_a \ (\mu g \ m^{-3}) = A \ c_w \ (ppm)$$

where A is an empirical constant with values varying from 0.03 to 0.2.

Keating and McKone (1992) measured the size distribution of particles larger than 1 μm in a shower (see also Section 5.4.2) for three different nozzle types. Unfortunately, no measurements of the submicrometer aerosol population were reported, and the effect of droplet shattering was not investigated. The importance of this submicrometer aerosol population (enhanced by the presence of a human in the shower) is poorly understood.

4.3.4.2   Aerosol Production by Humidifiers

Highsmith et al. (1988, 1992) used commercially available humidifiers (ultrasonic, impeller, and steam) with tap water containing approximately 300 mg L^{-1} of dissolved solids in a series of residences. Their preliminary experiments and their follow-up study indicated that use of typical tap water for home humidification can produce very high PM$_{10}$ (mass of particles with diameters less than 10 μm) concentration levels, as high as 7000 μg m^{-3} in an extreme case for a closed room and an ultrasonic humidifier. Whole-house PM$_{10}$ levels exceeding 40 μg m^{-3} were observed even if the ultrasonic humidifier was operated using a sample of commercially purchased distilled water as recommended by the manufacturer. Their data suggested that approximately 90% of the contents dissolved in the water are emitted into the indoor environment in the respirable fine fraction size range. Ultrasonic humidifiers, and to a lesser degree impeller humidifiers, are efficient aerosol generators for most of the dissolved impurities in water. Steam humidifiers, on the other hand, appear to produce no measurable increase in particulate levels. The authors suggested that the effect of ultrasonic humidifiers on total indoor aerosol concentration (c$_a$ in μg m^{-3}) can be predicted by

$$c_a = \frac{c_w \, \text{WCR} \, \text{F} \, 1000}{\text{V M} \, (\text{AER} + \text{K}_L)} + c_b$$

where c$_w$ is the water mineral content in mg L^{-1}, WCR is the humidifier water consumption rate in L h^{-1}, AER is the house or room exchange rate in h^{-1}, F is the proportion of particles generated in the specific size fraction, V is the volume of the room or house (in m^3), M is a unitless factor for the degree of mixing, K$_L$ is the loss rate of particles due to settling (in h^{-1}), and c$_b$ is the background particle concentration (in μg m^{-3}). Their study suggested values of K$_L$ of 0.14 h^{-1} for fine and 0.28 h^{-1} for coarse particles. Values of F for fine particles of 0.9 were reported for ultrasonic humidifiers.

Hardy et al. (1992) investigated the potential of portable humidifiers to produce asbestos fibers if used with asbestos-contaminated water. Their limited-scale experiments indicated that asbestos structures can be readily aerosolized by ultrasonic humidifiers with airborne concentrations being directly proportional to the charging water-asbestos concentrations. Similar tests conducted with an impeller humidifier resulted in asbestos concentrations varying from 3 to 18% of the corresponding ultrasonic humidifier experiments. The particles were mainly thin fibers, smaller than 6 μm in length. They concluded that indoor asbestos concentrations exceeding the ambient background asbestos concentration may occur if an ultrasonic humidifier, charged with asbestos-contaminated water, is operated in a closed room.

4.3.4.3   Evaporation of Volatile Chemicals from Droplets

Giardino et al. (1992) also conducted experiments in which trichloroethylene (TCE) was dissolved in the shower water, and the amount of volatilization of the TCE was measured during operation of the shower at room temperature. The average percentage of volatilization was 67% and 59% at 5 and 10 L min^{-1}, respectively. The difference was attributed in part to the different droplet size distributions at the two flow rates. It was determined on the basis of the original data that most of this volatilization occurred as the droplets fell to the bottom of the shower, but a significant fraction of the volatilization also may have occurred from the standing pool of water at the bottom of the shower. Overall, these results are consistent with those of McKone and Knezovich (1991), who found an average percentage of about 60% for TCE volatilization from both hot and cold showers.

Finally, Giardino et al. (1992) compared their experimental results with calculations of TCE volatilization from water droplets using three different mass-transfer models. The models overpredicted the amount of volatilization of the TCE by varying amounts. However, the trends predicted by two of the models agreed qualitatively with the results, namely that the 5 L min^{-1} droplet size distribution has a smaller transfer efficiency than the 10 L min^{-1} distribution. These results show that droplet size distributions can have an effect on exposure to volatilized gases as well as to the droplets themselves and that further study of their size distributions is necessary.

4.3.4.4 Summary

Little or no information exists regarding droplet and aerosol production during most indoor tap water uses. For example, water-related particle production during cooking, household cleaning, and toilet use have been neglected by practically all studies. An order of magnitude characterization of the strength of the various sources/activities outlined in Section 4.4.2 is necessary. The information that exists about showering is preliminary and relies heavily on a single experiment. The submicrometer aerosol distribution in a shower environment has not been characterized. More experimental studies similar to the one by Gunderson and Witham (1988) are necessary. Use of a mannequin is recommended, as most respirable particles are probably produced during the collision of the droplets with the body of the person in the shower. A better understanding of the interactions between droplet formation, evaporation, and removal is necessary. Our understanding of the production of aerosols during humidifier use has significantly improved. Ultrasonic humidifiers and to a lesser degree impeller humidifiers can be significant aerosol sources.

REFERENCES

Friedlander S. K. (1977) Smoke, Dust, and Haze: Fundamentals of Aerosol Behavior, John Wiley & Sons, New York.

Giardino, N. J., N. A. Esmen, and J. B. Andelman (1992) Modeling volatilization of trichloroethylene from a domestic shower spray: the role of drop-size distribution, Environ. Sci. Technol., 26, 1602–1606.

Gunderson E. C. and C. L. Witham (1988) Determination of inhalable nonvolatile salts in a shower, SRI International, SRI Project No. PYC-6631, Menlo Park, California.

Hardy R. J., V. R. Highsmith, D. L. Costa, and J. A. Krewer (1992) Indoor asbestos concentrations associated with the use of asbestos contaminated tap water in portable home humidifiers, Environ. Sci. Technol., 26, 680–689.

Highsmith N. R., C. E. Rodes, and R. J. Hardy (1988) Indoor particle concentrations associated with use of tap water in portable humidifiers, Environ. Sci. Technol., 22, 1109–1112.

Highsmith N. R., R. J. Hardy, D. L. Costa, and M. S. Germani (1992) Physical and chemical characterization of indoor aerosols resulting from the use of tap water in portable home humidifiers, Environ. Sci. Technol., 26, 673–679.

Hinds W. C. (1982) Aerosol Technology, John Wiley and Sons, New York.

Keating G. A. and T. E. McKone (1992) Measurements and evaluation of the water to air transfer and air concentration for trichloroethylene in a shower chamber, Pittsburgh, PA, NTIS/DE93007498.

McKone, T. E. and J. P. Knezovich (1991) The transfer of tricholoroethylene (TCE) from a shower to indoor air: experimental measurements and their implications, J. Air Waste Manag. Assoc., 41, 832–837.

Nazaroff, W. W. and G. R. Cass (1989) Mass transport aspects of pollutant removal at indoor surfaces, Environ. Int., 15, 567–584.

Nazaroff, W. W., L. G. Salmon, and G. R. Cass (1990a) Concentration and fate of airborne particles in museums, Environ. Sci. Technol., 24, 66–77.

Nazaroff, W. W., M. P. Ligocki, T. Ma, G. R. Cass (1990b) Particle deposition in museums. Comparison of modeling and measurement results, Aerosol Sci. Tech., 13, 332–348.

Nazaroff, W. W. and G. R. Cass (1991) Protecting museum collections from soiling due to the deposition of airborne particles, Atmos. Environ., 25, 841–852.

Owen M. K., D. S. Ensor, and L. E. Sparks (1992) Airborne particle sizes and sources found in indoor air, Atmos. Environ., 26A, 2149–2162.

Pruppacher H. R. and J. D. Klett (1980) Microphysics of Clouds and Precipitation, D. Reidel, Hingham, Mass.

Seinfeld, J. H. (1986) Air Pollution, John Wiley & Sons, New York.

Seinfeld J. H. and S. N. Pandis (1997) Atmospheric Chemistry and Physics: From Air Pollution to Global Change, John Wiley & Sons, New York.

4.4   WATER (DIRECT CONTACT)

Clifford P. Weisel

4.4.1   PRINCIPLES

Direct contact with residential tap water occurs primarily during activities associated with washing (showering, bathing, washing dishes), with a secondary exposure in the handling of wet items. For most people, showering and bathing are the major sources of dermal exposure to contaminants in water. Swimming and the use of hot tubs also may represent significant sources of exposure for some people. Skin absorption of contaminants as vapors or aerosols is generally insignificant compared to absorption directly from the water. Dermal exposure is a function of the duration of direct contact with water, the surface area of the skin contacting the water, and the water contaminant concentration. The internal dose resulting from the exposure depends on the permeability of each contaminant through the skin, as discussed in Chapter 6. Information on the frequency and duration of different washing activities for the U.S. population is presented in Chapter 3. Age and gender differences in some of these values have been observed; for example, the skin's surface area for different age groups and genders is well documented in the EPA *Exposure Factors Handbook* (1997).

4.4.2 Chemical Characteristics

A number of processes can alter the concentration of contaminants during various uses of household water compared to the concentration present in the water delivered to the home. These include: releases of volatile constituents when the water is exposed to air during showering, bathing, or opening the tap to obtain a rapid water flow; volatilization when the water is boiled for hot beverages and in cooking; production of additional DBPs by reactions of residual disinfectants or thermal destruction of labile compounds while the water is stored within the household hot water heater or during cooking; and removal of contaminants by home purification systems. Thus the concentration of the contaminant in the water during showering, bathing, and washing that directly contacts skin can differ from the concentration in the water delivered to the home. The transfer of volatile contaminants from water during showering is governed by the mass transfer of volatile species from the water to the air, which for many compounds is diffusion rate limited. A detailed discussion of the mathematical modeling of this process is presented in Section 4.2. The volatilization of compounds from water reduces their concentrations in the water that directly contacts the skin and therefore reduces the potential dermal exposure.

Experimental studies to determine the transfer efficiencies of volatile compounds during showering have been conducted by measuring the compounds' concentrations in the water leaving the shower head and at the shower drain or in the air (Giardino et al. 1988, McKone and Knezovich 1991, Tancrede et al. 1992, Giardino et al. 1992) (see Section 4.2). Transfer efficiency values ranging between 20 and 75%, have been measured for trichloropropane, trichloroethylene, chloroform, carbon tetrachloride, tetrachloroethylene, and radon. Water temperature, water flow rate, and the Henry's Law constant for the compound volatilized have been proposed as important controlling factors. Some of the volatilization of compounds occurs while the water is on the floor or walls of the shower stall, so transfer efficiencies calculated from experimental data in a shower stall would tend to underestimate the concentration of volatile compounds in the water contacting the body.

McKone and Knezovich (1991) mathematically modeled the release of trichloroethylene from water, based on an extrapolation of coefficients calculated for radon gas. These modeled data were evaluated by measuring the concentration in the inlet and effluent water using a laboratory shower. Based on their model, a mass transfer efficiency of 44% was calculated, with the mean of the experimental data being 61±9%. The authors reported that the transfer efficiency appears to depend upon the diffusion coefficient in water and does not have a strong temperature dependence for compounds with high, dimensionless Henry's Law constants (H > 0.1). The transfer efficiency could also be a function of the aerosol size distribution, since smaller aerosols have higher surface area to volume ratios than larger aerosols and the contaminants can diffuse to the surface more quickly. Thus, a greater quantity of volatile species will be released into the air from smaller-sized aerosols.

The impaction of water on the body, which results in a spreading of the water into a thin layer, could facilitate both the transfer of contaminants into the body and the volatilization of compounds from the water. It should be noted that Jo et al.

(1990) determined that water impaction on the body did not measurably alter the shower air concentration of chloroform compared to the shower air concentration without a person being present. This suggests that additional losses of volatile compounds from water while on the body are small compared to other volatilization processes that occur in the shower. A further consideration in predicting the concentration of volatile contaminants in the water contacting the body is the relationship between the fraction released and the distance that the water has fallen. This relationship has not been established. This relationship is important since the water contacts and remains on the body for different time intervals after leaving the shower head and equilibrium is not reached.

Thus, the concentration of volatile compounds in water in direct contact with the body is lower than the concentration leaving the shower head but higher than that exiting at the shower drain, the two locations where water concentration measurements have been made to determine the transfer efficiency. The transfer efficiency needs to be considered on a compound-specific basis. The concentration in water contacting the body could be as little as one fourth the original concentration for highly volatile compounds with low water solubility. Additional data are needed to refine this estimate. This includes determining the time profile of contaminant release within a shower stream, how long water stays on the body, and the release rate of volatile compounds from water contacting skin.

Measurements of volatilization of organic compounds directly from bath water are not available, though the release rate is expected to be less than from showering since the surface area to volume ratio of the water is smaller in a bathtub than in a shower. Releases of volatile compounds also occur during the filling of a bath. Estimates of the release of volatile compounds while water flows from a tap have been made for filling of kitchen sinks (see below).

Changes in contaminant concentration in water also occur in a hot water heater (Weisel and Chen 1994). Disinfection byproducts of chlorine are produced at an accelerated rate at the elevated temperatures in a hot water heater, continuing until the chlorine residual is completely consumed. Nearly complete consumption of the chlorine residual occurred within 8 hours at 65°C, with more than half the residual consumed within 30 minutes, times that are relevant to household water usage. The increases in the concentrations of the THMs and other thermally stable compounds are dependent upon the amount of chlorine residual and bromide ion present, which will vary within and between water distribution systems depend on the disinfection process, the season, and the residence time of the water in the pipes prior to reaching the residence (Krasner et al. 1989). Chlorine residuals of less than 1mg/L increased the THM concentration by about 20µg/L when water was heated (Weisel and Chen 1994). The proportions of the different chlorinated and brominated species produced were similar to those within the distribution system. Some compounds (trichloropropanone) decomposed when heated, while others initially increased due to formation by chlorine residual reactions and then later decreased due to thermal instability (dichloroacetolnitrile). Heating water is also known to increase the solubility of inorganic salts, increasing the concentration of dissolved metals and anions. Since showering, bathing, and other washing activities use a mixture of hot and cold water, estimates of contaminant concentrations for these activities should consider potential differences in the concentrations from the two sources.

Exposures from other washing activities that result in direct water contact, such as washing hands and dishes, are expected to be smaller than from showering or bathing, since the time involved and/or the amount of surface area of skin contacted are smaller. The contaminant water concentrations in sinks are reduced for volatile compounds compared to the concentrations leaving the tap (Howard and Corsi 1995). Volatile compounds are released when the water is sprayed from the faucet, with the presence of a bubble aerator increasing the release, and from volatilization from standing water in the sink or bath. Howard and Corsi (1995) determined that the amount released from spraying water was a function of the Henry's Law constant for three compounds having widely different constants, and that gas-phase resistance to transfer was dominant. Less than 5% of a soluble compound (acetone) and between 20 and 40% of a sparsely soluble compound (cyclohexane) was stripped from the water when a kitchen sink was used. These values are consistent with the transference observed in showers.

Additional activities that present potentially significant dermal exposure to water contaminants for a subset of the population are swimming and using a hot tub. However, the profile of contaminants in the swimming pool or hot tub is often quite different from that of the tap water used to fill the pool or hot tub, because the water usually remains in swimming pools or hot tubs for extended time periods and is sometimes filtered or is treated with additional disinfectants. Heating of the water can accelerate the loss of volatile and thermally/chemically labile water contaminants. Nonvolatile compounds will increase in concentration as water evaporates and is replaced. Filtering can remove some contaminants. Disinfectants added to the water will result in the production of disinfection byproducts. The dermal exposure associated with contaminants from the tap water used to fill the pools and tubs would therefore only be expected to be dominant for nonvolatile, stable contaminants not produced by the disinfectants added to the water. Swimming and hot tub use involves a more limited segment of the population than the other activities involving direct water contact and can have a seasonal component. These activities can represent a major pathway for direct contact with water contaminants for the portion the population that regularly swims (Beech et al. 1980, Aggazzotti et al. 1990). Swimming needs to be examined separately for children, since wading pools for young children may be filled on a daily basis without added disinfectants, the water may be different from that in larger swimming pools, and children may spend more time in recreational swimming than adults.

In summary, the parameters that appear to have the largest effect on dermal exposures to contaminants in water are those affecting the water concentration and the activities that result in dermal contact. The activities of major concern are the duration and frequency of the water contact during bathing, showering, and swimming. The main factors that affect the contaminant concentrations are the concentration in the water entering the home, emissions of the volatile compounds from water that decrease the water concentrations, and storage of water in hot water heaters prior to use. These processes can increase or decrease compound concentration depending on the thermal stability and volatility of the compound and the presence of a residual disinfection agent. In swimming pools and hot tubs, increases in nonvolatile compounds also can occur as water evaporated from the pool and is replaced by additional water.

4.4.3 KNOWLEDGE GAPS

A definitive characterization of how water contaminant concentrations change for volatile species during use, particularly showering, has yet to be developed. It should be possible to model and experimentally evaluate the amount of volatilization that occurs based on compound-specific data (such as Henry's Law constants), water temperature, water surface area or aerosol formation, and the time available for volatilization. Several previous studies calculating potential inhalation exposures have data applicable for estimating this release (see Chapter 4.2). Only limited data are available on changes in the water concentration of contaminants within a residence, with the exception of lead leaching from pipes and plumbing fittings. The percentage of the body that contacts water during showering and bathing, and whether the water concentration changes on the skin during showering, have not been reported. The importance of these effects on dermal absorption is discussed in detail in Chapter 6. The amount of time spent showering, bathing, swimming, and using hot tubs for high-exposure groups and potentially sensitive populations, such as children, has not been adequately determined.

REFERENCES

Aggazzotti, G., Fantuzzi, G., Righi, E., Tartoni, P., Cassinadri, T., and Predieri, G. (1990) Chloroform in alveolar air of individuals attending indoor swimming pools, *Archives of Environmental Health* 48, 250-254.

Becch, J. A., Diaz, R., Ordaz, C., and Palomeque, B. (1980) Nitrates, chlorates and trihalomethanes in swimming pool water, *American Journal of Public Health* 70, 79–81.

Bunge, A., Flynn, F. L., and Guy, R. H. (1994) Predictive Model for Dermal Exposure Assessment. In Wang, R. (Editor) *Water Contamination and Health*, Marcel Dekker, New York, pp 347.

Giardino, N. J., Esmen, N. A., and Andelman, J. B. (1992) Modeling Volatilization of Trichloroethylene from a Domestic Shower Spray: The Role of Drop Size Distribution, *Environmental Science Technology,* 26, 1602–1606.

Giardino, N., Gumerman, E., Andelman, J.,Wilkes, C., and Small, M. (1990) Real-time Air Measurements of Trichloroethylene in Domestic Bathrooms Using Contaminated Water. In Proc. 5th Int. Conf. Indoor Air Quality and Climate, Toronto, July 1990, 2, 707.

Howard, C., and Corsi, R. L. (1995) Volatilization of Chemicals from Drinking Water to Indoor Air: Role of the Kitchen Sink. *Journal of Air & Waste Management Association* 46, 830–837.

Jo, W., Weisel, C., and Lioy, P. (1990) Chloroform Exposure and the Health Risk Associated with Multiple Uses of Chlorinated Tap Water. *Risk Analysis* 10, 581–585.

Krasner, S. W., McGuire, M. J., Jacangelo, J. G., Patania, N. L., Reagen, K. M., and Aieta, E. M. (1989) The occurrence of disinfection by-products in U.S. drinking water. *Journal of the American Water Works Association* 81, 41–53.

Lindstrom, A., Highsmith, V., Buckley, T., Pate, W., and Michael, L. (1994) Gasoline-Contaminated Ground Water as a Source of Residential Benzene Exposure: A Case Study. *J. of Exposure Analysis and Environmental Epidemiology* 4, 183–195.

McKone, T. E., and Knezovich, J. P. (1991) The Transfer of Trichloroethylene (TCE) from a Shower to Indoor Air: Experimental Measurements and Their Implications. *Journal of Air & Waste Management Association* 41, 832–837.

Tancrede, M., Yanagisawa, Y., and Wilson, R. (1992) Volatilization of Volatile Organic Compounds from Showers — I. Analytical Method and Quantitative Assessment.*Atmospheric Environment* 26A, 1103–1111.

U.S. Environmental Protection Agency (EPA), (1997) *Exposure Factors Handbook,* EPA/600/P-95/002Fc, U.S. EPA, Washington, D.C.

Weisel, C. P., and Chen, W. J. (1994) Exposure to Chlorination By-Products from Hot Water Uses, *Risk Analysis* 14, 101–106.

4.5 MODELING OF EXPOSURE TO WATERBORNE CONTAMINANTS

Charles R. Wilkes

4.5.1 INTRODUCTION

Estimating a person's exposure to waterborne contaminants requires the consideration of uptake via three pathways: (1) respiratory, (2) dermal, and (3) ingestion. It also requires the consideration of the contaminant concentration in the water or air during each contact event, along with the frequency, duration, and nature of the contact. A multitude of factors influence the ultimate exposure to an individual, including physical characteristics of the environment, behavior of the individual with respect to the contaminant, and chemical properties of the contaminant. This section provides an overview of the modeling topics and concepts that should be considered when conducting a model-based exposure assessment. For an illustrative application of these concepts, refer to the case study.

The most common, cost-effective method of estimating human exposure to airborne contaminants (the inhalation route) is through indoor air quality and exposure modeling. In addition to the reduced cost, exposure modeling techniques provide several advantages over field studies. A field study provides a single-point estimate of an exposure for a given sequence of events. Modeling an exposure event can provide considerably more information. For example, using Monte Carlo-type simulation techniques, exposure distributions can be estimated for the numerous combinations of activity patterns and environmental conditions. From these distributions one can analyze the higher exposures in the distribution to determine the types of activities and environments that lead to these exposures. Also, higher-risk population subgroups can be identified.

The ultimate goal of exposure/risk modeling is to estimate exposure/risk to an individual or population resulting from the presence of a chemical in the environment with sufficient accuracy to provide the knowledge base to make informed decisions on how to protect public health. This can be accomplished in a variety of ways, including stochastic models, deterministic models, and combined stochastic and deterministic models.

Purely stochastic models represent the exposed concentrations by sampling from a distribution of observed concentrations in a similar environment. This technique is the simplest, but requires much data to achieve a degree of accuracy. Data must be collected specific to the population group being analyzed to represent the range of situations that may be encountered. A significant drawback to this approach is the disconnection between cause and effect. Events that have a large impact, particularly on the upper extreme of the exposure distribution, may be inadequately represented. For example, a person taking a shower immediately after a previous shower is exposed to significantly higher air concentrations than an individual who uses a shower stall that has not been used in several hours.

Purely deterministic models are useful for simulating a specific case of interest. Deterministic techniques typically utilize mass-balance models, solving sets of differential equations to predict concentrations and exposure. These models idealize the environment as a collection of well-mixed zones interconnected by flows (Axley, 1988; Sandberg, 1984; Sinden, 1978) and mathematically represent important processes leading to contaminant generation, transport, and removal. These types of models require a reasonably complete understanding of all significant factors impacting concentrations and exposure. When conducting a population exposure assessment, defining the parameters on an individual basis for each member of a population is unrealistic.

For population exposure assessments, a combined stochastic and deterministic model provides the ability to represent the variability inherent in a population group while maintaining the necessary connection between cause and effect. This approach samples input parameters for a deterministic model from distributions based on estimated values, or samples input parameters from a database of measured or observed values. This approach is recommended for estimating the distribution of exposures to a population.

4.5.2 Modeling Methods

Implementing a combined stochastic and deterministic model for estimating the distribution of exposures to a population group can be achieved by defining the deterministic model framework and sampling appropriate input parameters from databases or representative distributions. The deterministic model is defined to represent the physical environment for a set of sampled parameters and estimates one point in the distribution of exposures. Deterministic models are typically based on mass conservation, and implement a mass balance calculation around a collection of assumed well-mixed zones. Figure 4.5.1 illustrates an arbitrary system of n zones connected by one-directional flow elements. A zone is defined as a well-mixed air parcel. Zones, also referred to as compartments, are typically determined by the physical boundaries within a building, and can be made up of individual rooms, collections of rooms, or the entire building. Additionally, multiple zones can be used to represent a room that is not well mixed. Flow elements represent the transport processes from one compartment to another.

Another technique that was considered but not used is the use of microenvironments (or single compartments). The use of microenvironments is a method of

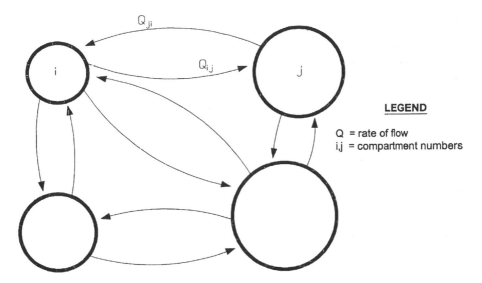

FIGURE 4.5.1 Idealization of a Building's Flow-System (after Sinden, 1978). The circles represent zones; the lines with arrows represent flow elements.

simplifying the modeling task while retaining most of the reality of the indoor environment. Duan (1982) defined microenvironments as "a chunk of air space with homogeneous pollutant concentration." Microenvironments are idealized representations of the various indoor locations encountered by an individual during normal activities. Rooms or buildings that the individual occupies at the time of interest are considered microenvironments. The use of microenvironments allows a stochastic representation of the important events that occur within the microenvironment without requiring that the activities of all individuals passing through the microenvironment be accounted for explicitly.

When the individual enters a microenvironment, an initial air concentration is assigned based on survey data for that type of room. The pollutant-generating activities of the individual are explicitly modeled while the individual occupies the microenvironment. In addition, activities of other individuals are either ignored or attributed as characteristics of the microenvironment. This is a significant drawback because it does not represent recent events in that microenvironment that would impact air concentration. For example, if one person takes a shower immediately following another person, the significant initial air concentrations experienced by the second person are neglected. A shortcoming of the use of microenvironments is the artificial separation of the microenvironments. When a pollutant-generating activity occurs in a microenvironment, its effect is not seen in other microenvironments (i.e., other rooms in the same house).

In contrast, the method we used is an aggregate technique involving the use of one or more "representative" environments for frequently occupied spaces, with other less-common environments represented by microenvironments. To represent the home, a common house configuration can be defined as the "representative" environment. A "representative" environment is one made up of multiple zones possibly inhabited by multiple individuals. Each zone (or room) may impact the other zones via flow paths, and each individual impacts the other individuals through their activities. Particular parameters, such as room volumes and air exchange rates can be sampled from representative distributions. A similar approach can be used to represent other frequent or important environments, such as the workplace. For example, a population of office workers may be depicted as a "representative" residence and office building, while other minor environments, such as a grocery store, may be depicted as a microenvironment.

4.5.2.1 General Case

A building is represented as a collection of well-mixed zones, modeling the characteristic volumes, airflows, sources, and sinks. Mass-balance equations about the compartment boundary are formulated, and these ordinary differential equations are solved using numerical techniques.

The mass-balance differential equation for a compartmental representation is:

$$V_i \frac{dC_i}{dt} = \sum_{j=1}^{n} q_{j,i} C_j - \sum_{j=1}^{n} q_{i,j} C_i + \sum sources \pm \sum sinks \qquad (4.5.1)$$

where: C = air concentration in the compartment ($\mu g/m^3$)
V = volume of the compartment (m^3)
t = time (hours)
q = volumetric air flowrate entering or leaving the compartment (m^3/hour)
i,j = compartment numbers
n = total number of compartments
sources represent generation mechanisms present in the compartment
sinks represent removal mechanisms (both reversible and irreversible sinks) present in the compartment

The source strength for VOCs volatilizing during water uses has been shown to be well described by the two-film or dual resistance representation (Andelman et al., 1987, Giardino et al., 1992, Wilkes et al., 1992). Substituting the two-film representation for the source generation rate and neglecting sink effects yields the following equation:

$$V_i \frac{dC_i}{dt} = \sum_{j=1}^{n} q_{i,j} C_j - \sum_{j=1}^{n} q_{j,i} C_i + \sum_{g=1}^{m} k_g F_g \left(C_{w0} - \frac{C_i}{H} \right) \qquad (4.5.2)$$

where: g = source number in compartment i
m = total number of sources in compartment i
k_g = volatilization coefficient for source g, ($0 \le k_g \le 1.0$, unitless)
F_g = volumetric water flowrate of source g (m^3/hour)

C_{wo} = concentration of the contaminant in the water supply prior to volatilization (μg/m^3)

H = Henry's Law constant (unitless)

The above differential equation cannot be solved for the general case, and therefore numerical solution techniques are typically used. Each of the parameters can be set or sampled from representative databases or distributions based on measured values for that type of environment.

4.5.2.2 Single Compartment Case (Microenvironment)

To calculate concentrations using the microenvironment (one-compartment) simplification, assuming $q = q_{out} = q_{in}$ with the initial condition $C(t = 0) = Co$, Equation 4.5.2 is integrated to yield the following equation:

$$C(t) = \left[\frac{q}{V} C_b \left(1 - \exp(-zt)\right) + \frac{\sum_{g=1}^{m} k_g F_g}{V} C_{wo} \left(1 - \exp(-zt)\right) + z C_o \exp(-zt) \right] * \frac{1}{z} \quad (4.5.3)$$

where: C_b = background concentration (exterior) (μg/m^3).

C_o = initial concentration in the microenvironment (μg/m^3).

q = volumetric airflow rate entering and leaving the microenvironment (m^3/hour).

$$z = \frac{q}{V} + \frac{\sum_{g=1}^{m} k_g F_g}{VH}$$

4.5.2.3 Water Uses

Volatile organic compounds (VOCs) in the water supply enter the indoor air during normal water uses, resulting in inhalation exposure. This inhalation exposure often exceeds the exposure that occurs through ingestion of water from the same source, with showers providing a significant contribution (Andelman, 1985; Jo et al., 1990a; Jo et al., 1990b; McKone, 1987; Wilkes et al., 1992). The mass transfer of the volatile compounds from the water supply is represented as discussed in Section 4.2.

The term *source clouds* has been used to represent the higher concentration in the air near the source. This higher concentration area has been shown to be important when the individual is in the proximity of the source, as occurs during an active use (Rodes et al., 1991; Rector et al., 1990; Koontz et al., 1988), and can have a significant impact on the potential inhaled dose to an individual. Although the source cloud is likely to be an important factor in the ultimate exposure level, not enough information is currently available to incorporate this effect in modeling efforts.

For the purpose of modeling, water uses can be classified as passive uses (uses that occur whether or not the individual is present) and active uses (uses that the individual initiates). Passive water uses occur automatically as a result of activities of others or as a result of automatic equipment (i.e., a dishwasher that is started by a timer). The distinction between passive and active events is important, primarily due to the importance of the proximity of the modeled individual to the source.

Passive water uses can be accounted for in either a representative environment or in a microenvironment by assigning water-use events to the environment. These events can be described deterministically or stochastically. For example, if the environment is a kitchen with a dishwasher, then one dishwasher event per day may be assigned to the environment to account for a family's usage. The environment might also be assigned characteristic faucet uses to represent the uses of family members other than the modeled individual. The water uses by the modeled individual (active uses) would be defined separately based on the characteristics of that individual.

4.5.2.4   Initialization

The concentration in an environment with moderate to frequent water uses is likely to be greater than zero when any individual enters, due to prior water-use activities by the individual as well as water-use activities by others. To attain a better estimate of exposure, the effect of prior water uses can be considered by simulating water-use activities that occurred during an appropriate period prior to entry of the individual. The period of time prior to entry that needs to be considered is dependent primarily on the air exchange rate of the compartment. The time for the microenvironment to undergo three complete air exchanges is adequate time to effectively eliminate any residual concentrations from prior events. Since air exchange rates are typically greater than about 0.2 ACH, the time period that needs to be considered is usually less than about 15 hours. Alternatively, an initial concentration can be sampled from a representative distribution.

4.5.2.5   Exposure Calculations

Human exposure is generally defined as contact at the boundary between a human and a chemical, physical, or biological agent, and quantified as the concentration of the agent in the contact media integrated over the duration of the contact. (USEPA, 1997, Glossary). Exposure to VOCs originating in multiple environments (e.g., residence bathroom, residence kitchen, office building, and school) can be modeled by allowing the activity pattern of the individual to dictate the individual's location, and subsequent contact with the chemical.

The potential inhaled dose is defined as the amount of the chemical in the air passing into the body. The resulting calculation of potential inhaled dose is an integration over time of the individual's breathing rate multiplied by the concentration time-series present at the body envelope. The resultant calculated potential dose represents the amount of the contaminant that enters through the breathing zone and is available for uptake (USEPA, 1992; Lioy, 1990):

$$PID = \int C_a(t) * BR(t) * dt \qquad (4.5.4)$$

where: *PID* = potential inhalation dose (i.e., the mass of the contaminant that
              enters the lungs, μg).
        $C_a(t)$ = air concentration at the location of the individual as a function
              of time (μg/m³).
        $BR(t)$ = breathing rate as a function of time (m³/hour).

Refer to Chapter 5 for a complete discussion of respiratory uptake. A dermal exposure event can be characterized by the amount of contact area, the length of contact, and the concentration of the chemical in the media. Refer to Chapter 6 for a complete discussion of dermal uptake.

4.5.3   REPRESENTATION OF INPUT PARAMETERS

The various input parameters described above have considerable variability across the population and housing stock. Activity patterns have been shown to have a very large impact on the ultimate exposure to an individual (Wilkes et al., 1996). Other factors, including air exchange rates, building and room volumes, water consumption rates, and water concentrations also have a significant impact on exposure, and need to be carefully characterized.

Activity patterns are either predetermined by an individual or scenario of interest or are sampled from a representative database. The activity pattern databases currently available are described in Chapter 3 and are sampled to establish the location of the modeled individual within the modeled environment. Since water uses of other individuals in the environment impact concentrations, these people need to be accounted for by either modeling the other occupants or by attributing these uses to the environment.

Air exchange rates are primarily a function of the housing type and climate, which varies with geography. Refer to Chapter 3 for information on ranges of house and room volumes and ventilation rates.

The contaminant concentration is a characteristic of a local water supply and needs to be supplied for the scenario of interest. When information is available, parameter values should be sampled from a distribution reflecting the actual range of variability to ensure a reasonable estimate of the distribution of exposures to a population.

4.5.4   KNOWLEDGE GAPS

The impacts of the model inputs used for representing microenvironments and other environments are understood to varying degrees. As a result of numerous field studies, the range of variability in air exchange rates is generally understood for selected environments based on the building characteristics and the location's prevailing weather patterns. The source behavior of volatilization from the water supply during showers is also generally understood as a result of prior research (Andelman,

1985; Giardino et al., 1992; Little, 1992). However, little research has been conducted on sources other than showers.

The interaction of these processes with human behavior leads to an even greater uncertainty in estimating the distributions of exposures. Human behavior affects the air exchange rate through opening and closing of doors and windows, the source strength through water use, the potential for exposure through location patterns, and the exposure rate through breathing and physical contact with the water. Research has estimated that variability in human behavior can lead to at least a tenfold difference in potential inhaled dose as a result of volatilization of VOCs from the water supply (Wilkes et al., 1996).

Published studies validating the microenvironment modeling framework are rare. While the use of microenvironments is potentially an effective and discerning technique for quantifying exposures of population groups, the validation of these techniques requires extensive field studies. The distribution of exposures to given population groups would need to be quantified and compared to modeling efforts to allow a meaningful interpretation of the validity of these techniques.

REFERENCES

Andelman, J.B., Inhalation exposure in the home to volatile organic contaminants of drinking water, *The Science of the Total Environment,* 47, 443–460, 1985.

Andelman, J.B., Wilder, L.C., and Myers, S.M., Indoor air pollution from volatile chemicals in water, *Proceedings of the 4th International Conference on Indoor Air Quality and Climate,* Berlin, 37–41, August 1987.

Axley, J., Progress toward a general analytical method for predicting indoor air pollution in buildings, *Indoor Air Quality Modeling Phase III Report,* NBSIR 88-3814, National Bureau of Standards, National Engineering Laboratory, Center for Building Technology, Building Environment Division, July 1988.

Duan, N., Models for human exposure to air pollution, *Environment International,* 8, 305–309, 1982.

Giardino, N.J., Gummerman, E., Esmen, N.A., Andelman, J.B., Wilkes, C.R., Davidson, C.I., and Small, M.J., Shower volatilization exposures in homes using tap water contaminated with trichloroethylene, *Journal of Exposure Analysis and Environmental Epidemiology,* Suppl. 1, 147–158, 1992.

Jo, W.K., Weisel, C.P., and Lioy, P.J., Routes of chloroform exposure and body burden from showering with chlorinated tap water, *Risk Analysis,* 10(4), 575–580, 1990a.

Jo, W.K., Weisel, C.P., and Lioy, P.J., Chloroform exposure and the health risk associated with the multiple uses of chlorinated tap water, *Risk Analysis,* 10(4), 581–585, 1990b.

Koontz, M.D., Rector, H.R., Fortmann, R.C., and Nagda, N.L. Preliminary Experiments in a Research House to Investigate Contaminant Migration in Indoor Air, United States Environmental Protection Agency, EPA 560/5-88-004, June 1988.

Lioy, P.J., Assessing total human exposure to contaminants, *Environmental Science & Technology,* 24, 938–994, 1990.

Little, J.C., Applying the two-resistance theory to contaminant volatilization in showers, *Environmental Science & Technology,* 26(7), 1341–1349, 1992.

McKone, T.E., Human exposure to volatile organic compounds in household tap water: the indoor inhalation pathway, *Environmental Science & Technology,* 21, 1194–1201, 1987.

National Academy of Sciences (NAS), Human exposure assessment for airborne pollutants: advances and opportunities, National Academy Press, Washington, D.C., 1991.

Rector, H., Koontz, M., Nagda, N., and Kennedy P., Determination of exposures using contaminant migration patterns, *Proceedings — The Fifth International Conference on Indoor Air Quality and Climate*, Toronto, Canada, 2, 577–582, July 1990.

Rodes, C.E., Kamens, R.M., and Wiener, R.W., The significance and characteristics of the personal activity cloud on exposure assessment measurements for indoor contaminants, *Indoor Air*, 2, 123–145, 1991.

Sandberg M., The multi-chamber theory reconsidered from the viewpoint of air quality studies, *Building and Environment*, 19(4), 221–233, 1984.

Sinden F.W., Multi-chamber theory of air infiltration, *Building and Environment*, 13, 21–28, 1978.

U.S. Environmental Protection Agency (USEPA), *Exposure Factors Handbook*, EPA/600/P-95/002Fc, Washington, D.C. 1997.

U.S. Environmental Protection Agency (USEPA), Guidelines for Exposure Assessment, *Federal Register*, 57(54), 22888–22938, 1992.

Wilkes C.R., Modeling human inhalation exposure to VOCs due to volatilization from a contaminated water supply, Ph.D. Dissertation, Department of Civil Engineering, Carnegie-Mellon University, Pittsburgh, PA, April 1994.

Wilkes C.R., Small M.J., Andelman J.B., Giardino, N.J., and Marshall J., Inhalation exposure model for volatile chemicals from indoor uses of water, *Atmospheric Environment*, 26A(12), 2227–2236, August 1992.

Wilkes C.R., Small M.J., Davidson C.I., Andelman J.B., Modeling the effects of water usage and co-behavior on inhalation exposures to contaminants volatilized from household water, *Journal of Exposure Analysis and Environmental Epidemiology*, 6(4), 393–412, 1996.

5 Respiratory Uptake

Robert R. Mercer

5.1 INTRODUCTION

Potential routes for entry of drinking water contaminants include the gastrointestinal, dermal, and respiratory systems. Understanding the relative contribution of each of these routes to the uptake of toxicants is critical in assessing the health risks from contact with drinking water contaminants. Human activities that involve contact with drinking water include the air-phase release of volatile drinking water contaminants and/or direct contact such as drinking, bathing, food preparation, cooking, and washing.

While the respiratory system uptakes of a variety of gas and particle toxicants have been extensively characterized for human exposures due to environmental pollutants (Miller et al., 1982; Overton, 1984; Schlesinger, 1985; Gargas and Andersen, 1988), the methods developed for these cases have not generally been applied to situations relevant to uptake of drinking water contaminants or the comparison of dermal and inhalation routes. In this respect the dermal route of absorption has been more extensively studied. However, by the nature of its function in gas exchange the human respiratory track is likely to be a permissive barrier for uptake of airborne drinking water toxicants. In one day the process of respiration in a normal adult brings in over 22 m³ of air into close contact with a 300 m² alveolar surface that is perfused by essentially all of the cardiac output (Weibel, 1983). By comparison, the dermal barrier has a surface area of 2 m² and is perfused with less than 5% of cardiac output (Rao and Brown, 1993).

1-57881-001-9/99/$0.00+$.50
© 1999 by International Life Sciences Institute

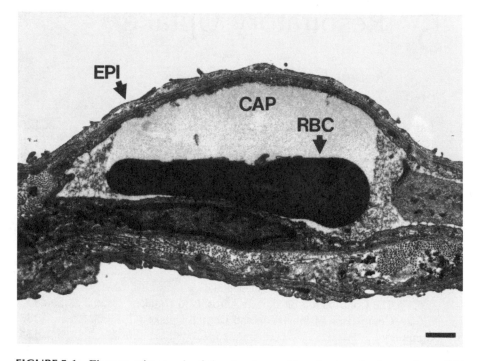

FIGURE 5.1 Electron micrograph of the alveolar-capillary barrier of a human lung. EPI epithelial cell layer, CAP capillary lumen, RBC red blood cell (bar = 1 μm).

5.2 REGIONS OF THE RESPIRATORY TRACT

The design of the lungs is optimized for uptake of oxygen and simultaneous removal of carbon dioxide (Weibel, 1983). The exchange of oxygen and carbon dioxide is accomplished by lung structures that are optimized for efficient transport. The cellular, anatomic, and three-dimensional architecture of this design play a critical role in assessing the uptake of inhaled toxicants as described in comprehensive texts on the subject of inhalation toxicology (Gehr and Crapo, 1988).

The transport pathway for gas exchange is conveniently divided into three functional regions. These are the upper respiratory tract, which includes both nasal and mouth passages for breathing of air; the tracheobronchial or conducting airways, which include the airways from trachea to the smallest, nonrespiratory airways; and the alveolar lined ducts, which form the gas exchange region of the lungs. Gas exchange across the alveolar–capillary barrier is accomplished by a large three-dimensional surface in which the air-to-blood barrier for gas diffusion is only about 0.2 μm thick, as illustrated in Figure 5.1 (Stone et al., 1992).

The upper respiratory tract and conducting airways protect the fragile gas exchange surface by conditioning the humidity and temperature of inspired air and by filtering and rapid clearance of inhaled particles and bacteria. The upper respiratory tract and conducting airways are covered by a viscoelastic mucous layer

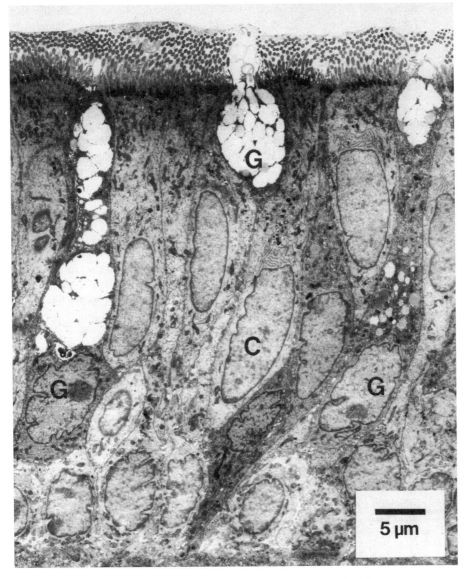

FIGURE 5.2　Electron micrograph of mucous layer, cilia and epithelial cells of a human airway. G, Goblet cell; C, Ciliated cell (bar = 5 μm).

(Figure 5.2). The mucous layer tapers in thickness from the upper respiratory tract (10 μm) to the most distal conducting airways (2 μm) (Mercer et al., 1991). The beating of cilia from the ciliated cells of the airway epithelium constantly propels the mucous layer upward to the glottis, where the mucous is removed by swallowing. This so-called mucociliary escalator provides a rapid and efficient means for removal of inhaled particles and bacteria. Clearance from the gas-exchange region is known to be much slower, as no such physical conveyor is present. However, each alveolus of the human lung contains an average of 12 macrophages, which are thought to process all particles reaching the gas-exchange region under normal conditions (Stone et al., 1992).

5.3 MODELS FOR ASSESSING RESPIRATORY DOSIMETRY

Because the physical mechanisms of transport and chemical uptake vary significantly between different airborne pollutants, no single method of approach can be used to estimate the pulmonary uptake of all inhaled toxicants. For instance, highly reactive, but low-tissue-solubility gaseous toxicants such as nitrogen dioxide and ozone are not uniformly distributed within the air spaces of the lung. These pollutants, which are typical of irritant gases, primarily react with the mucous lining layer in the upper respiratory tract and thus do not have a significant concentration in the systemic circulation. Assessing the respiratory dosimetry of these airborne pollutants requires a model approach, which accounts for differences in concentration gradients in different regions and how the concentration gradients vary at different times during each breath (Miller et al., 1982, Overton, 1984).

The majority of toxicants present in drinking water have a relatively low tissue metabolism/reactivity and thus a relatively uniform distribution within the lungs. Most drinking water contaminants of interest rapidly diffuse across the pulmonary–capillary barrier (DuBois and Rogers, 1968). Pulmonary uptake of these toxicants is controlled by ventilation/perfusion rates, blood/air partition coefficients and the concentration gradient across the pulmonary capillary barrier. Once delivered to the systemic circulation, toxicants may cause toxic reactions in other organs by direct action of the parent chemical or its toxic metabolites. Elimination of toxicants in the systemic circulation is chiefly due to metabolism or excretion.

There are three basic approaches used in models to assess respiratory dosimetry. These are:

1. The *fractional extraction* approach, in which an empirically determined constant is used to estimate the fraction of the inspired dose that is absorbed. This approach has the virtue that the uptake is accurately known for the toxicant under the specific conditions of the study. These conditions include the subject's weight, metabolism, ventilation rate, and tidal volume. The method is limited by the need to conduct studies in the human subjects and, due to its empirical approach, must be conducted separately for every toxicant even when the toxicants are chemically similar.
2. In the *maximal absorbed dose* approach the air to blood solubility or Henry's Law constant is used to estimate the fraction of the inhaled toxicant that is absorbed. According to Henry's Law the solubility of a gas phase compound in an adjacent liquid phase is proportional to the partial pressure of the toxicant in the gas phase and a proportionality constant referred to as the Henry's Law constant. In this approach the loss of toxicants due to excretion and/or metabolism is generally neglected and the concentration or back pressure in the blood and/or tissue barriers

is assumed to be zero. As a result of neglecting these factors, which would otherwise reduce the estimate of uptake, the method provides an estimate of the maximally absorbed dose. The method has the advantage of being readily applicable to the study of a variety of toxicants where the Henry's Law constant can be directly measured by simple means or can be estimated by the air-to-water partitions coefficient for the compound or a related species.

In the ideal case, one would also take into consideration the transport across the various compartments of the tissue phase. These barriers to transport would include the liquid lining layer to epithelial cell layer, interstitial matrix and cell layer and the endothelial cell layer. For many inhaled toxicants including the potential effects of each barrier is not necessary, as the lungs are optimally designed for air-to-blood transport via a large surface area and thin tissue barrier. Because of this optimal design the lungs do not significantly impede the transport of low-molecular-weight compounds. Most drinking water contaminants that are of concern for health risk evaluation are low-molecular-weight compounds and are thus easily transported across the lungs when inhaled. Larger chemicals with low lipid solubility cannot easily penetrate the lipid membranes of cells and must pass the junctional spaces between cells of the lungs. However, the cells forming the epithelium beneath the liquid lining layer and the cells forming the endothelium encircling the capillary are interconnected by tight junctions. The tight junctions form tortuous, narrow aqueous passages between cells that prove relatively impenetrable to chemicals with molecular weights greater than approximately 1000 Daltons.

3. In the *physiologically based pharmacokinetic* (PBPK) modeling approach the uptake of toxicants is determined by a mass balance approach that takes into account uptake by the lungs, metabolism in body tissues, and excretion. In the PBPK model a series of differential equations are solved, which include the physiological parameters of lung ventilation, cardiac output, and blood exchange of systemic tissues with pharmacokinetic parameters of tissue metabolism/excretion, and air/tissue partition coefficients of the toxicant. The drinking water toxicants that are evaluated in this report have a relatively low lung tissue metabolism/reactivity and can thus be modeled by the PBPK approach.

Figure 5.3 shows the structure of a representative PBPK model that includes the contribution of drinking water contaminants for uptake by respiratory systems as an aerosol or gas and uptake by the dermis. As in the model of Chinery and Gleason (1993), the skin is broken into the outer, nonperfused layer, or stratum corneum, and an inner perfused layer.

Even a relatively minimal PBPK model requires measurements of numerous constants such as ventilation rates, the Henry's Law constant, and metabolized fraction of the toxicant, blood flow, and organ volumes. Due to the large number of factors in the model analysis of the sensitivity of the model results in variability in each factor is an important element in validation of model predictions. As shown

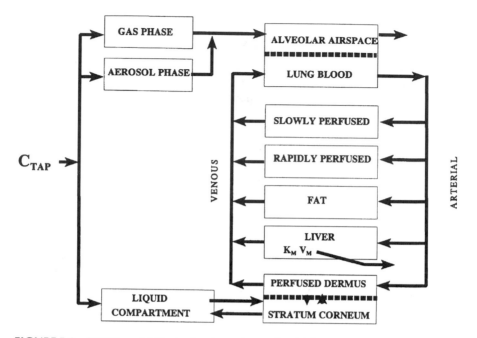

FIGURE 5.3   PBPK model illustrating the pathways for drinking water toxicant uptake and metabolism.

in theoretical studies (Cox, 1995), the PBPK approach is basically a linear, coupled system of differential equations whose various parameters ultimately reflect a single equivalent parameter. Ideally, the PBPK model should provide the theoretically based method to obtain an estimate of the same factional extraction coefficient that is directly measured in the empirically based fractional extraction approach.

In addition to the gas phase route of entry, the uptake of aerosols by toxicants released during water use may constitute an important route for uptake of drinking water contaminants. Deposition of aerosols in the lungs occurs according to physical mechanisms of inertial impaction, gravitational sedimentation, and diffusion (Yeh et al., 1976; Lippmann, 1977; Willeke and Baron, 1993). A variety of factors such as aerosol particle size, density, shape, hygroscopic/hydrophobic character, and electrostatic charge may play important roles in determining the location and efficiency of deposition in the lungs. Because particles are present in a range of sizes, an aerosol is typically described by a size distribution or a mass/count weighted mean. In toxicologic evaluations the mass median aerodynamic diameter (MMAD) is most frequently used to describe the aerosol, as it reflects both the aerodynamic diameter and delivered mass of particles.

Both empirical and mathematical approaches have been used to assess the dosimetry of the inhaled particle (Lippmann, 1977; Schlesinger, 1985). Direct measurements of deposition have demonstrated that the human upper respiratory tract

TABLE 5.1
Characterization of Shower Generated Aerosols

Aerosol Diameter μm	Aerosol Concentration #/cm³	Aerosol Mass ng/cm³
1.3	180–660	0.2–0.7
2.1	90–320	0.4–1.4
3.4	40–120	0.8–2.3
5.5	20–60	1.6–4.8
Total	360–1200	2.9–9.2

efficiently removes particles greater than approximately 5 μm MMAD (Schlesinger, 1985). For particles in the 1 to 5 μm range, the total respiratory tract deposition (upper respiratory+conducting+gas exchange regions) efficiencies for particles are on the order of 20%. Mathematical model-based estimates of the alveolar deposition efficiency of inhaled 1 and 5 μm aerosol particles have been estimated to be 5.2 and 17%, respectively.

5.4 SOURCES OF AIRBORNE DRINKING WATER TOXICANTS

5.4.1 VOLATILE GASES

Shower and faucet use result in the release of volatile drinking water contaminants such as radon and volatile organic compounds into the air phase. The release of organic chemicals from the drops of shower spray is dependent on the concentration gradient between air and liquid phases, temperature and Henry's Law constant for the compound (McKone and Knezovich, 1991). The Henry's Law constant is a measure of the relative solubility of the chemical between gas and air phases. Experimental studies have demonstrated that 40 to 70% of volatile organic toxicants, such as trichloroethylene, are released into the air phase during showers (Giardino et al., 1992; Keatin and McKone, 1992).

5.4.2 AEROSOLS

Secondary to the production of a spray of large, nonrespirable water droplets, showers produce a distribution of respirable liquid aerosols. Keatin and McKone (1992) measured the aerosol size distribution (Table 5.1) and volatilization of trichloroethylene produced by showers. The volume of the shower simulation chamber was 1.05 m³ with an airflow of 65 L/min. Two standard nozzles (4.2 and 6.0 L/min) and one water-saving nozzle (2.8 L/min) were examined. Though droplet size of the standard nozzles was significantly larger (1000 to 1500 μm) than that of the water-saving nozzle (300 μm) there was no significant difference in the size distribution of aerosols produced. Volume mean diameters ranged from 7.1 to 16.3 μm with respective aerosol concentrations of 300 and 1200 cm³.

TABLE 5.2
Age-Activity Respiratory Parameters

Age (years)	Sedentary Activity Frequency (Breaths per minute)	Sedentary Activity Tidal Volume (milliliters)	Maximal Activity Frequency (Breaths per minute)	Maximal Activity Tidal Volume (milliliters)
0.6	35	42	70	165
1.8	28	84	70	328
4.0	22	152	68	570
8.2	18	266	60	1190
20.0	14	500	40	3050

5.4.3 OTHER SOURCES

In addition to the aerosols generated from water uses described in Chapter 4, there are many sources of indoor and ambient aerosols that humans are exposed to. Examples include showers, hot-water faucets, and humidifiers. These nondrinking water sources of aerosols may provide a significant mechanism for inhalation exposure of volatilized drinking water toxicant if the toxicant significantly binds to the ambient aerosol. For instance, attachment of radon progeny to submicron particles and the subsequent deposition of these particles in the lungs is the primary mechanism by which radon-induced alpha particle irradiation of the lungs occurs (Harley and Pasternack, 1972). In the home, nondrinking water sources of aerosols include burning candles, cigarettes, vacuuming, drying clothes, cooking, and contributions from outside ambient air. Hopke et al. (1990) measured the distribution of submicron particles (0.5 to 500 nm size range) in a home using a condensation nucleus counter. Particle concentrations were 5500 to 10,000 cm^3 in the bedrooms, 9000 to 13,000 cm^3 in the living and dining rooms, and 80,000 to 100,000 cm^3 in the kitchen. The high particle concentrations in the kitchen were attributed to a gas pilot light in the stove. Presumably gas pilot lights in other sources such as forced air heaters and water heaters would produce similar high particle concentrations.

5.5 PHYSIOLOGIC AND PHARMACOKINETIC FACTORS

5.5.1 RESPIRATORY PARAMETERS

Standard nomograms may be used to determine how resting respiratory parameters of tidal volume and breathing frequency vary with age and/or body weight (Comroe, 1974). There are potentially significant differences in human respiratory uptake due to variability in individual respiratory parameters. A large portion of these variations in respiratory parameters are due to differences in age and activity patterns. These effects have been examined in models designed to assess the age-dependent sensitivity in adverse health effects following exposure to inhaled pollutants. Table 5.2 gives the age-dependent changes in respiratory parameters for both sedentary and maximal activity derived by Hofmann et al. (1989) for studies of particle deposition.

The data demonstrate that frequency of ventilation decreases during development whereas tidal volume increases at a greater rate, resulting in a much higher respiratory minute volume. These data are specifically formulated by the International Commission on Radiation Protection in order to predict the particle dosimetry of a normal adult male (ICRP,1975). Gender-dependent differences in respiratory parameters are known to be normalized by body weight scaling, which can be used to adjust the respiratory parameters for comparison of male and female differences in toxicant uptake.

5.5.2   Other Factors

Demographic factors significantly affect the daily and lifetime exposure of individual. These factors include population mobility and mortality, activity patterns at work and home, and the characteristics of the building and water supply/mitigation. Inclusion of these factors is done in a two-step process that is described in the Exposure Characteristics and Exposure Estimates sections of this chapter. In the first phase the exposure concentrations are estimated in each area or microenvironment of each building where human activity occurs. In the second phase the activity patterns of individuals are used to determine the degree of exposure in each of the microenvironments.

5.6   MODEL ESTIMATES OF LUNG TOXICANT UPTAKE

5.6.1   The Contribution of Shower Aerosols to Uptake

Based on the efficiency of aerosol deposition in the lungs and the shower aerosol concentrations given in Table 5.1, it is possible to estimate the mass of shower aerosol deposited in the human lung due to a 30-minute shower at standard breathing conditions of 500 ml tidal volume and a breathing rate of 30 breaths per minute. These calculations show that between 20 and 60 μg of water and water-soluble, nonvolatile matter contained in the aerosol phase are deposited on the lung surface due to showering. This level of uptake is insignificant compared to the uptake due to gas-phase transport determined from PBPK models. The results demonstrate that shower aerosols do not have the potential to deliver significant doses of nonvolatile, soluble drinking water contaminants to the lungs.

It should be noted that this conclusion does not necessarily apply to insoluble contaminants, which may not have significant gas-to-tissue phase diffusion across the dermal and respiratory barriers. For instance, respirable aerosols generated from showers and hot-water faucets have been shown to trap microorganisms and may play a significant role in disease transmission (Bollin et al., 1985, Castellani et al., 1988). As yet, no insoluble chemical contaminants have been identified.

5.6.2   Comparison of Inhalation and Dermal Uptake

Mathematical models such as physiologically based pharmacokinetic models (PBPK) have been used to estimate the relative contribution of skin and inhalation adsorption due to showering for tetrachloroethylene (Rao and Brown, 1993) and chloroform (Chinery and Leason, 1993). Inhalation for tetrachloroethylene was estimated to deliver 92 to 93% of the total concentration delivered to the brain, whereas inhalation delivery of chloroform was expected to deliver approximately 57% of the total. Uncertainties in the air–skin and air–blood partition coefficients used for these two compounds appear to be the principal factor in these disparate results.

5.7 SUMMARY

- There are considerable differences in the relative contributions of dermal vs. inhalation uptake of drinking water contaminants from published studies.
- Aerosols generated by showering are not a significant route for uptake of nonvolatile, water-soluble drinking water contaminants.

REFERENCES

Bollin, G. E., Plouffe J. F., Para, M. F., and Hackman, B., Aerosols containing "Legionella pneumophilia" generated by shower heads and hot-water faucets. *Appl. Environ. Microbiol.*, 50,1128, 1985.

Castellani Pastoris, M., Vigano, E. F., and Passi, C., A family cluster of legionella pneumophila infections. *Scand. J. Infect. Dis.*, 20, 489, 1988.

Chinery, R. L. and Leason, A. K., A compartmental model for the prediction of breath concentration and absorbed dose of chloroform after exposure while showering. *Risk Anal.*, 13, 51, 1993.

Comroe, J. H., *Physiology of Respiration.* Year Book Medical Publishers, Chicago, 1974, Chap. 3.

Cox, L. A., Simple relations between administered and internal doses in compartment flow models. *Risk Anal.*, 15, 197, 1995.

DuBois, A. B. and Rogers, R. M., Respiratory factors determining the tissue concentration of inhaled toxic substances. *Respir. Physiol.*, 5, 34, 1968.

Gargas, M. L. and Andersen M. E., Physiologically based approaches for examining the pharmacokinetics of inhaled vapors. *Toxicology of the Lung.* Gardner, D. E., Crapo, J. D. and Massaro, E. J., Eds., Raven Press, New York, 1988, 449.

Gehr, P. and Crapo, J. D., Morphometric analysis of the gas exchange region of the lung. *Toxicology of the Lung.* Gardner, D. E., Crapo, J. D., and Massaro, E. J., Eds., Raven Press, New York, 1988, 1.

Giardino, N. J., Esmen, N. A., and Andelman, J. B., Modeling volatilization of trichloroethylene from a domestic shower spray: The role of drop-size distribution. *Environ. Sci. Tech.* 26, 1602, 1992.

Harley, N. H. and Pasternack, B. S., Alpha absorption measurements applied to lung dose from radon daughters. *Health Phys.* 23, 771, 1972.

Hofmann, W., Martonen, T. B., and Menache, M. G., Age-dependent lung dosimetry of radon progeny. *Extrapolation of Dosimetric Relationships for Inhaled Particles and Gases.* Crapo, J. D., Miller, F. J., Smolko, E. D., Graham, J. A., and Hayes, A. W., Eds., Academic Press, New York, 1989, 317.

Hopke, P. K., Li, C. S., and Ramamurthi, M., Air cleaning and radon decay product mitigation. *Hanford symposium on health and the environment: Indoor radon and lung cancer — reality or myth (29th).*, NTIS/DE93007318, Richland, WA, 1990, 479.

ICRP, Task Group on Lung Dynamics, Deposition and retention models for internal dosimetry of the human respiratory tract. *Health Phys.*, 12, 173, 1966.

Keatin, G. A. and McKone, T. E., Measurements and evaluation of the water-to-air transfer and air concentration for trichloroethylene in a shower chamber. *Symposium on modeling of indoor air quality and exposure*, Pittsburgh, PA, NTIS/DE93007498, 1992, 1.

Lippmann, M., Regional deposition of particles in the human respiratory tract. *Handbook of Physiology.*, American Physiology Society, Bethesda, MD, 1977, 213.

Mckone, T. E. and Knezovich, J. P., The transfer of trichloroethylene (TCE) from a shower to indoor air: Experimental measurements and their implications. *J. Air Waste Mange. Assoc.* 40, 282, 1991.

Mercer, R. R., Anjilvel, S., Miller, F. J., and Crapo, J. D., Inhomogeneity of ventilatory unit volume and its effects on reactive gas uptake. *J. Appl. Physiol.* 70, 2193, 1991.

Mercer, R. R. and Crapo, J. D., Three-dimensional analysis of lung structure and its application to pulmonary dosimetry models. *Toxicology of the Lung.* Gardner, D. E., Crapo, J. D., and McClellan, R. O., Eds., Raven Press, New York, 1993, 155.

Mercer, R. R., Russell, M. L., and Crapo, J. D., Radon dosimetry based on the depth distribution of nuclei in human and rat lungs. *Health Phys.* 61,117, 1991.

Miller, F. J., Anjilvel, S., Menache, M. G., Asgharian, B., and Gerrity, T. R., Dosimetric issues relating to particulate toxicity. *Inhalation Toxicol.* 7, 615, 1982.

Miller, F. J., Overton, J. H., Meyers, E. T., and Graham, J.A., Pulmonary dosimetry of nitrogen dioxide in animals and man. *Air Pollution by Nitrogen Oxides.* Schneider, T. and Grant, L., Eds., Elsevier Science, New York, 1982, 377.

Overton, J. H., Physiochemical processes and the formulation of dosimetry models.*J. Toxicol. Environ. Health,* 13, 273, 1984.

Rao, H. V. and Brown, D. R., A physiologically based pharmacokinetic assessment of tetrachloroethylene in groundwater for a bathing and showering determination. *Risk Anal.,* 13, 37, 1993.

Schlesinger, R. B., Comparative deposition of inhaled aerosols in experimental animals and humans. *J. Toxicol. Environ. Health* 15, 197, 1985.

Stone, K. C., Mercer, R. R., Gehr, P., Stockstill, D., and Crapo, J. D., Allometric relationships of cell numbers and size in the mammalian lung. *Am. J. Respir. Cell Mol. Biol.* 6, 235, 1992.

Weibel, E. R., Is the lung built reasonably?, *Am. Rev. Respir. Dis.* 128, 752, 1983.

Willeke, K. and Baron, P. A., *Aerosol Measurement Principles, Techniques, and Applications.* Van Nostrand Reinhold, New York, 1993.

Yeh, H. C., Phalen, R. F., and Raabe, O. G., Factors influencing the deposition of inhaled particles. *Environ. Health Perspec.* 15, 147, 1976.

6 Dermal Uptake

Annette L. Bunge and James N. McDougal

CONTENTS

1-57881-001-9/99/$0.00+$.50
© 1999 by International Life Sciences Institute

6.1   INTRODUCTION

Mammalian skin acts as the interface between the unfriendly environment outside and the carefully controlled aqueous milieu inside. It is a complex organ with protective, structural, sensory, and thermoregulatory functions. The skin makes up about 10% of total body weight. A primary protective function of the skin under normal conditions is to prevent excessive loss of body water. The physical characteristics of the skin that limit the flux of water are found in the tightly packed, lipid–protein matrix on the surface of the skin — the stratum corneum. The stratum corneum keeps us from drying out like raisins in the air and from swelling like balloons in fresh water.

The skin, however, is not a complete barrier to water or any chemical. Many chemicals can penetrate the skin well enough to cause systemic toxicity, and the skin is actually used as a route of entry with several transdermal drug delivery devices. The primary characteristics of chemicals that determine how well they will penetrate the skin are molecular size, molecular charge, and solubility. The physical form in which a chemical is presented to the skin also influences the dermal penetration rate by affecting the amount of chemical that is available for penetration (i.e., its thermodynamic activity). For example, a chemical in water is usually more available for dermal penetration than when the same chemical is present at the same concentration in soil.

Skin contact with contaminated drinking water comes primarily through bathing and showering, although other activities such as washing dishes, clothes, and children or cleaning house can provide dermal exposures. While few actual measurements of the absorbed dose following showering or bathing have been made, it has been demonstrated that penetration of contaminants does occur (Jo et al. 1990; Weisel and Shepard 1994).

6.1.1   EXPOSURE TO VAPORS

As detailed in Section 4.2, during showering and bathing household water contaminated with volatile organic chemicals can provide a vapor exposure to skin in addition to the primary inhalation exposure. Organic chemical vapors have been demonstrated to be absorbed through the skin of human volunteers (Dutkiewicz and Piotrowski 1961, Gromiec and Piotrowski 1984, Riihimaki and Pfaffli et al. 1978, Wieczorek 1985), monkeys (Hefner et al. 1975, McCollister et al. 1951), rats (McDougal et al. 1985), and nude mice (Tsuruta 1989). Several *in vitro* studies have also measured organic chemical vapors penetration through isolated skin (Barry et al. 1984, Blank and McAuliffe 1985, Jacobs and Phanprasit 1993). Generally, penetration of chemical vapors through the skin would be expected to contribute 10% or less to the internal dose in a whole body exposure where inhalation also occurs (McDougal et al. 1990). All of these studies have focused on vapor exposure to dry skin; there is no information about the affect of a water film on the skin during a vapor exposure. Dermal absorption of chemical from direct exposure to vapors during showering and bathing is considered negligible.

6.1.2 Exposure to Aerosols

As described in Section 4.3, aerosols can be generated during showering. There are no known studies that have studied purely aerosol exposures to the skin. Aerosol droplets would be expected to coalesce into a film of water on the skin. In addition, if the chemical dissolved in the aerosol droplet is more volatile than water, the process of aerosol formation and coalescence will promote loss of chemical from the aerosol. As a result, the concentration of the volatile component in the coalesced water on the skin would be less than the concentration in water that was not aerosolized. If the chemical in the aerosol is less volatile than water, evaporation of water from the aerosol droplet will increase the concentration of chemical. However, dermal absorption from chemicals dissolved in aerosol droplets is expected to be small relative to that from liquid water, primarily because aerosols will contact less skin area than liquid water. For these reasons, dermal absorption from aerosols during showering and bathing is considered negligible.

6.1.3 Exposure to Liquid Water

Exposure to contaminated water by shower and bath water is considered the primary mechanism by which chemicals are presented to the skin. Penetration of chemicals from aqueous solutions has been widely studied in humans and laboratory animals, both *in vivo* and *in vitro*. These studies are too numerous to mention individually and interested readers are encouraged to examine compilations by Leung and Paustenbach (1994), McKone and Howd (1992), Flynn (1990), the U.S. EPA (1992), and Vecchia (1997). Dermal absorption is commonly described by the permeability coefficient, which represents quantitatively the ease with which a chemical penetrates the skin. Permeability coefficients reported for organic chemicals from aqueous solutions range from 5.5×10^{-6} cm/hr^{-1}(nitrosodiethanolamine) to 1.2 (ethyl benzene) (U.S. EPA, 1992). Pure water at 25°C has a permeability coefficient of about 0.5×10^{-3} to 1.5×10^{-3} cm/hr^{-1} through human skin (Blank 1953, Bronaugh 1986). Approximately 90% of the chemicals that have permeability coefficients measured from aqueous solutions have higher permeability coefficients than water. Given sufficient duration and extent of exposure, the potential for significant dermal absorption of chemicals from contaminated water exists.

6.2 BARRIER PROPERTIES

6.2.1 Fick's Law

Skin barrier function has been extensively studied through *in vitro* and *in vivo* experiments on a variety of animal skins including human. These experiments provide compelling evidence that penetration through skin is reasonably represented as diffusion through a pseudo-homogeneous membrane as described by Fick's first law. Fick's law states that the flux of the penetrating chemical at a location within the membrane barrier is proportional to the membrane diffusion coefficient and the

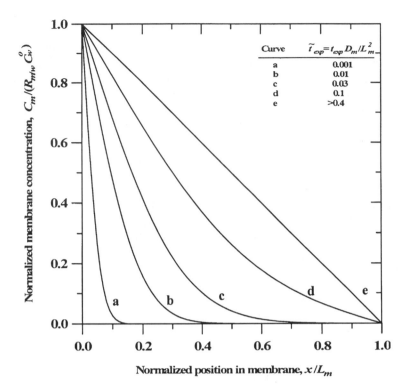

FIGURE 6.1   Normalized concentration of chemical in the skin membrane as a function of time.

concentration gradient (i.e., the driving force) at that position. Insight into the process of dermal penetration can be gathered from mathematical solutions of Fick's law combined with a differential mass balance written for typical exposure scenarios.

Commonly, skin exposed to contaminated water would be free of chemical prior to exposure. Shortly after a chemical exposure begins, the chemical will penetrate a short distance into the skin membrane. As the exposure continues, the chemical penetrates further, to eventually permeate across the entire skin membrane. This situation is illustrated in Figure 6.1, which plots the chemical concentration in the skin (normalized by the maximum concentration, $R_{m/w}\, C_w^o$) as a function of the normalized distance into the skin (x/L_m) at different exposure times (t_{exp}) normalized by the characteristic time for diffusion across the skin (i.e., L_m^2/D_m). These curves were computed from Fick's law assuming that the skin behaves as a single membrane, and the chemical concentration (C_w) in the contaminated water did not change during the exposure (i.e., $C_w = C_w^o$ for the entire time of exposure), and that buildup of chemical in the systemic circulation was negligible (Crank 1975). In Figure 6.1, L_m is the apparent thickness of the skin membrane barrier, D_m is the apparent diffusion coefficient of chemical in the skin membrane, and $R_{m/w}$ is the equilibrium partition coefficient between the skin membrane and the contaminated water.

Figure 6.2 illustrates the effect of exposure time on the cumulative mass absorbed into the skin (M_{in}), and the cumulative mass leaving the skin to enter the body's

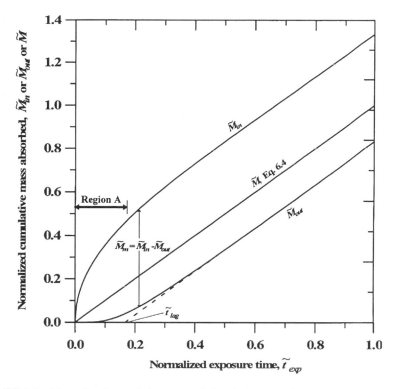

FIGURE 6.2 Normalized cumulative mass of chemical penetrating into and out of the skin membrane as a function of normalized exposure time.

interior system (M_{out}), for the case shown in Figure 6.1 (i.e., the body's interior system acts as an infinite sink). Both M_{in} and M_{out} are normalized by the theoretical maximum mass of chemical the skin can hold (i.e., $\tilde{M}_{in} = M_{in}/\left(R_{m/w}C_w^o L_m A\right)$ and $\tilde{M}_{out} = M_{out}/\left(R_{m/w}C_w^o L_m A\right)$ where A is the area of skin exposed to contaminated water. At a given exposure time, the difference between the $\tilde{M}_{in}$ and $\tilde{M}_{out}$ curves represents the normalized mass residing in the skin at that time. Whenever the exposure ends, the mass in the skin at that time ($\tilde{M}_{in} - \tilde{M}_{out}$) will continue to diffuse across the skin to enter the systemic circulation and will evaporate from the surface if the chemical is volatile. Theoretical calculations indicate that even for highly volatile chemicals, at least 1/3 of the chemical residing in the skin at the end of the exposure will continue to enter the skin. (This estimate was calculated by assuming that a steady-state concentration profile was established before evaporation began, and that the evaporation rate from skin was controlled by diffusion through skin.) For less volatile chemicals, most of the chemical residing in the skin when the exposure ends will eventually become systemically available.

Absorption is more rapid early in the exposure, when the chemical primarily fills the skin reservoir (i.e., the chemical has penetrated through only a portion of the skin membrane). During this time, almost no chemical leaves the skin to enter the systemic circulation. As the exposure continues, the chemical penetrates farther into the skin, causing the absorption rate to slow as the driving force decreases. If the exposure continues long enough, the chemical penetrates across the entire skin membrane to appear in the systemic circulation (i.e., M_{out} becomes larger than zero). If the exposure continues longer still, the concentration gradient through the membrane becomes constant. As a result, the rates at which mass enters and leaves the skin become equal and constant, as indicated by the linear portions of the normalized M_{in} and M_{out} curves in Figure 6.2. At this special condition, known as *steady-state* (*ss*),

$$\frac{dM_{in}^{ss}}{dt} = \frac{dM_{out}^{ss}}{dt} = A\,P_m\,C_w^o \tag{6.1}$$

where P_m is the permeability coefficient representing the chemical's ability to penetrate across the skin membrane. P_m should only be measured after steady state has been established. Measurement of P_m in a diffusion cell, which is not at steady state, based on the amount of chemical appearing in the receptor solution will give a value of P_m that changes with time. When measured at steady state, P_m is theoretically related to the membrane properties as (Crank 1975):

$$P_m = D_m\,R_{m/w}\big/L_m \tag{6.2}$$

For general comparisons of one chemical with another, only the time-invariant (i.e., the steady-state) permeability coefficients are useful. For this reason, correlations developed to predict permeability coefficients from molecular and physical property data (e.g., representations of molecular size and lipophilic-hydrophilic character) should be based on steady-state data.

The lag time for penetration across a homogeneous membrane ($t_{lag,m}$) is determined by extrapolating the linear portion (i.e., the steady-state part) of the M_{out} curve to zero (Figure 6.2). Theoretically, $t_{lag,m}$ is related to the skin membrane properties as (Crank 1975):

$$t_{lag,m} = \frac{L_m^2}{6\,D_m} \tag{6.3}$$

The length of time required to reach steady state is approximately 2.4 to 3 times $t_{lag,m}$ (Crank 1975).

Several documents have recommended estimating dermal penetration using (e.g., Leung and Paustenbach 1994, Walker et al. 1996):

$$M = A\,P_m\,C_w^o\,t_{exp} \tag{6.4}$$

where t_{exp} is the duration of the exposure. These documents usually do not explicitly state whether M is the mass that enters the skin or the mass that leaves the skin to enter the systemic circulation, although documents recommending Equation (6.4) often claim that it will overestimate the amount of dermal penetration because it does not consider the lag time. This equation may also underestimate the amount of penetration because it does not account for chemical remaining in the skin.

Equation (6.4) is plotted in Figure 6.2 as the normalized mass $\tilde{M}(= M / (R_{m/w} C_w^o L_m A))$. For a given exposure time, $\tilde{M}$ calculated from Equation (6.4) is always less than the total amount that would have entered the skin (i.e., $\tilde{M}_{in}$). During that same exposure period, not all of the chemical that has entered the skin will have been absorbed into the systemic circulation. Hence, $\tilde{M}$ is larger than the amount of chemical that has permeated all the way across the skin (i.e., $\tilde{M}_{out}$). However, after the exposure ends, the mass of chemical that has entered but not yet left the skin will continue to be systemically absorbed. For nonvolatile chemicals, evaporation from skin will be insignificant and the total amount of chemical that is eventually absorbed systemically as a result of an exposure of duration t_{exp} is better represented by M_{in} than by M. As already described, for highly volatile chemicals, about one third of the chemical in the skin at the end of the exposure may be absorbed systemically; the other two thirds may evaporate from the skin. Hence, even for highly volatile chemicals, M underestimates the amount of systemic exposure from dermal penetration by any amount of chemical remaining in the skin that will eventually penetrate. Consequently, for estimating the amount of dermal penetration into systemic circulation, M_{in} is more protective of human health than M.

Figures 6.1 and 6.2 illustrate absorption into and penetration across a membrane barrier with only one layer. However, the dermal barrier separating the capillary network from the outside environment consists of two main external layers, the stratum corneum and the viable epidermis. Skin structure and its impact on dermal penetration is discussed next.

6.2.2 Skin Structure

The effectiveness of skin as a barrier results primarily from the structural differences of the stratum corneum and the viable epidermis. Described simplistically, the outermost layer, the stratum corneum (*sc*), is composed of dead, essentially impermeable keratinized cells, surrounded by more permeable semicrystalline lipid bilayers. Beneath the *sc* lies the viable epidermis (*ve*), a layer of proliferating nucleated cells, which is easily recognized as the glistening (but not bloody) surface under a peeled blister. The capillary network lies beneath the *ve*, and chemicals that have diffused across the *ve* are then free to enter the systemic circulation. Several metabolizing enzymes are found in the esterase rich *ve*, and metabolism is known to occur to varying degrees for susceptible chemicals as they diffuse through the *ve*.

The lipophilic character of the *sc* minimizes penetration of hydrophilic chemicals like water, while the tortuous diffusion path around the dead cells delays and slows penetration of lipophilic chemicals that otherwise absorb readily into the lipid bilayers. The hydrophilic *ve* further attenuates the penetration rates of highly lipophilic chemicals.

Appendageal structures such as hair follicles, sebaceous ducts, and sweat glands may provide significant penetration pathways for ionizing chemicals (which are effectively blocked from passing through the lipid bilayers of the *sc*), and larger molecules ($MW > 350$ or so), which penetrate the *sc* slowly, especially during shorter exposures ($t_{exp} < t_{lag,m}$). Generally, this contribution is not significant because of the relatively small area fraction they represent (less than ~ 1 to 2% of the total area for abdominal human skin) (Scheuplein 1967a, Scheuplein and Blank 1971, Schaefer et al. 1982). The quantitative contributions of these routes are largely unknown since experiments measuring the rate and extent of absorption are not adequately sensitive to quantitatively differentiate between appendageal and dermal penetration. As a preliminary estimate, however, the permeability coefficient for a chemical transferring by this pathway may be calculated as 10% of D_w/L_{sc}, where D_w is the diffusion coefficient of the penetrating chemical in water.

When both the *sc* and *ve* contribute significantly to the skin's barrier, then the apparent permeabiltiy of the *sc–ve* composite membrane (P_m) is related in turn to the individual permeability coefficients of the *sc* (P_{sc}) and *ve* (P_{ve}):

$$\frac{1}{P_m} = \frac{1}{P_{sc}} + \frac{1}{P_{ve}} \qquad (6.5)$$

Equation (6.5) states that resistances (the inverse of permeability) in series are additive. Based on Fick's law, the steady-state permeability coefficients across the *sc* and the *ve* from an aqueous solution depend on the diffusion coefficients of the chemical in each skin layer (D_{sc} and D_{ve}), the diffusion path lengths of each (L_{sc} and L_{ve}), and the equilibrium partition coefficients of the chemical between the *sc* and water ($R_{sc/w}$) and between the *ve* and water ($R_{ve/w} = R_{ve/sc}/R_{sc/w}$):

$$P_{sc} = \frac{R_{sc/w} \, D_{sc}}{L_{sc}} \quad \text{and} \quad P_{ve} = \frac{R_{ve/w} \, D_{ve}}{L_{ve}} \qquad (6.6)$$

where $R_{i/w}$ is defined as the concentration of chemical in the skin layer *i* (mass/volume of membrane *i* at absorbing conditions) divided by the equilibrium chemical concentration of an aqueous solution (mass/volume) (Parry et al. 1990). The *sc* thickness varies with region, but is commonly reported as between 10 and 40 μm. By using transepidermal water loss data, Kalia et al. (1996a) measured 12.7 ± 3.3 μm for the *sc* of the ventral forearm of three subjects. The *ve* is approximately 50 to 100 μm thick (Scheuplein and Blank, 1971).

Functionally, lipophilic chemicals partition with difficulty into the relatively hydrophilic *ve*, while partitioning readily into the lipophilic *sc* (i.e., $R_{sc/w} > R_{ve/w} \cong 1$). However, because D_{ve}/L_{ve} is ~ 100 to 10,000 times larger than D_{sc}/L_{sc} (Scheuplein and Blank 1971), P_{ve} is so much larger than P_{sc} that $P_m \approx P_{sc}$, except for very lipophilic chemicals (with octanol-water partition coefficients larger than ~ 5000; that is, $\log_{10} K_{o/w} > ~ 3.7$) (Cleek and Bunge 1993, Bunge and Cleek 1995, Kasting and Robinson 1993, Vecchia 1997). Most chemicals appearing in drinking water will meet this condition, and any additional resistance from the *ve* can be neglected. In addition, if significant metabolism of the penetrating chemical occurs in the *ve*, the apparent permeability (i.e., the total penetration rate of parent and metabolite) through the *ve* will be larger than without metabolism. Consequently, if metabolism occurs, the sc may present the primary resistance across the skin barrier (i.e., $P_m \approx P_{sc}$) even if the penetrating chemical is more lipophilic than $\log_{10} K_{o/w} > \sim 3.7$. Dermal absorption of chemicals that arc mctabolizcd in thc ve may still bc a conccrn if thc mctabolitc(s) are possible health hazards themselves.

The lag time ($t_{lag,m}$) to cross the sc–ve composite membrane is related to the individual lag times for the sc ($t_{lag,sc}$) and the ve ($t_{lag,ve}$):

$$t_{lag,m} = t_{lag,sc}\left(\frac{3P_{sc} + P_{ve}}{P_{sc} + P_{ve}}\right) + t_{lag,ve}\left(\frac{P_{sc} + 3P_{ve}}{P_{sc} + P_{ve}}\right) \qquad (6.7)$$

where

$$t_{lag,sc} = \frac{L_{sc}^2}{6\,D_{sc}} \quad \text{and} \quad t_{lag,ve} = \frac{L_{ve}^2}{6\,D_{ve}} \qquad (6.8)$$

Consequently, if the sc is undamaged and $\log_{10}K_{o/w}$ for the absorbing chemical is less than about 3.7, then $P_{sc}/P_{ve} \ll 1$, $t_{lag,sc}/t_{lag,ve} \approx 10$ to 1000, and $t_{lag,m} = t_{lag,sc}$. The contribution of $t_{lag,ve}$ and P_{ve} become important for highly lipophilic chemicals (i.e., $\log_{10}K_{o/w}$ $>\sim 3.7$) or when the sc barrier is compromised by chemical or physical damage. Note that when the sc and ve layers both contribute significantly to $t_{lag,m}$ (i.e., the skin barrier is not a homogenous barrier), $t_{lag,m} \neq L_m^2/(6\,D_m)$ if L_m and D_m are the apparent thickness and diffusion coefficient, respectively, for the sc–ve composite. In addition, $P_m \neq R_{m/w}$ D_m/L_m if $R_{m/w}$ represents the partition coefficient between the sc–ve composite and water.

6.2.3 REGIONAL VARIATION

Skin from various regions of the body may differ both in thickness and in the density of appendages such as hair follicles, sweat ducts, and sebaceous glands. Measurements reported by Roy and Flynn (1990) indicate that there is no statistical difference in permeability measurements made using skin from the thigh compared to measurements from the abdomen for fentanyl and sufentanil. For other regions, however, variations in permeability of as much as an order of magnitude are reported (Scheuplein and Blank 1971, Maibach et al. 1971, Feldman and Maibach 1967).

Table 6.1 summarizes *in vitro* flux data for hydrocortisone (Feldman and Maibach 1967), parathion (Maibach et al. 1971) and water (Scheuplein and Blank 1971) normalized by flux through the ventral forearm. Hydrocortisone and parathion were applied to the skin using a small quantity of acetone. Generally, regions with a thinner sc or more appendages (e.g., the scrotum and scalp) exhibited higher fluxes. The notable exceptions are the palmar and plantar skin, which, according to Scheuplein, exhibited much greater water penetration than the forearm. By contrast, parathion and hydrocortisone penetration through palmar and plantar skin showed either a small or no enhancement relative to the forearm.

Differences in hydration may explain the different fluxes of hydrocortisone, parathion, and water through palmar and plantar skin relative to the ventral forearm.

TABLE 6.1
Flux through Skin from Various Body Regions Normalized by Flux through the Ventral Forearm for Hydrocortisone (Feldman and Maibach 1967), Parathion (Maibach et al. 1971) and Water (Scheuplein and Blank 1971)

Skin Region	Water	Parathion[†]	Hydrocortisone[†]
Abdomen	1.1	2.1	na*
Forehead	2.7	4.2	6.0
Scrotum	5.5	11.8	42.0
Back of Hand	1.8	2.4	na
Scalp	na	3.7	3.5
Plantar (sole)	12.6	1.6	na
Palm	3.7	1.6	0.8

* Not available
[†] Deposited on skin using acetone

Because hydrocortisone and parathion fluxes were determined by depositing chemical onto skin samples from acetone, which evaporates, the skin was probably not fully hydrated. However, water flux was measured in fully hydrated skin.

Scheuplein and Blank (1971) argued that the thicker, callused skin of the palms and soles is an inferior barrier to chemical penetration, at least when hydrated. If so, hydration may enhance penetration through some regions (e.g., palmar or plantar skin) more than through other regions (e.g., the forearm). For estimating chemical penetration from contaminated water, it may be more reasonable to assume regional variations of fully hydrated skin (i.e., use the water data in Table 6.1). However, if water enhances penetration, it may be more important for hydrophilic chemicals than for lipophilic chemicals, leading fully hydrated permeability coefficients to produce regional variations that depend on $K_{o/w}$. More studies are needed that examine regional variability of permeability from aqueous solutions for chemicals with a range of $K_{o/w}$ values. Based on present information, regional variations are not large enough to justify separate estimates of dermal absorption for various regions.

6.2.4 EFFECT OF SKIN DAMAGE

Skin condition can have a significant impact on the rate of penetration of chemicals, when the barrier function is disrupted (see reviews by Dugard and Scott 1984, Scheuplein and Bronaugh 1983). Bowman and Wahlberg (1989) conducted an excellent study in guinea pigs that showed that the effect of skin damage on blood levels is opposite for hydrophilic and hydrophobic chemicals. They showed that absorption of butanol (a hydrophilic solvent; $\log_{10}K_{o/w}$ partition coefficient of 0.65) is markedly increased by acute mechanical and chemical injuries, and the absorption of 1,1,1-trichloroethylene and toluene (hydrophobic solvents; $\log K_{o/w}$ of 2.49 and 2.73, respectively) is decreased by the same injury. In the case of butanol, Bowman and Wahlberg (1989) postulate that the main absorption barrier (i.e., the *sc*) is degraded and that in the case of the hydrophobic solvents edema induced by the damage may provide a secondary partitioning barrier. When the *sc* is compromised by chemical or mechanical damage, the barrier function of the *ve* (the P_{ve} and $t_{lag,ve}$) may be important. Except for individuals with extensive wounds or a chronic skin disease that significantly reduces barrier effectiveness, special consideration of damaged skin is probably unnecessary.

6.2.5 EFFECT OF TEMPERATURE AND BLOOD FLOW RATE

Skin temperature can have an impact on the rate of penetration of chemicals in two different ways. First, increasing the temperature of the skin has been shown to increase the rate of penetration by a direct effect in the skin that follows an Arrhenius relationship (Scheuplein and Blank 1971). Blank et al. (1967) showed that the steady state permeability coefficient of a series of alcohols through human epidermal membranes could be increased by 25 to 60% with a 5°C temperature change. These authors did not state how many hours were required to achieve steady state. Because of the nature of these *in vitro* studies, specifically the requirement that they be conducted long enough to reach steady state, they have little direct relevance to human showering and bathing exposures. Chang and Riviere (1991) showed that increasing the temperature by 5°C (38 to 42°C) doubled the flux of parathion through pig skin in a diffusion cell, but only after the temperature was elevated for several hours. A study of the effect of temperature on waterborne organic compounds (Jetzer et al. 1988) concludes "... little regard has to be taken of temperature when estimating the percutaneous absorption of pollutants from water and projecting the risks associated therewith."

Second, temperature may affect the blood flow to the skin and, therefore, affect the amount of chemical absorbed. Riviere and Williams (1992) suggest that blood flow to the skin can change by a factor of 100 depending on the thermal conditions. Whether or not altered blood flow affects absorption depends on the rate at which the chemical penetrates the skin. Poorly absorbed chemicals are less likely to be affected by changes in blood flow because absorption across the membrane is the rate-limiting factor. Rapidly absorbed chemicals are much more likely to be affected by blood flow. Auclair et al. (1991) showed that a 60% decrease in blood flow to a skin flap in hairless rats did not affect the absorption of urea but did decrease the systemic absorption of caffeine and progesterone. Siddiqui et al. (1989) showed that diffusional resistance accounts for about 99% of the absorption of steroids (i.e., blood flow accounts for only 1% of the resistance). Recent unpublished results (Gordon et al., 1997) suggest that exposure to cold water may decrease systemic absorption of a chemical, presumably by reducing cutaneous blood flow.

There has been little research on the effect of blood flow on penetration of organic chemicals from aqueous solutions. However, theoretical arguments suggest that changes in blood flow will have little effect on dermal penetration if the cutaneous blood flow per area of skin (i.e., volume of blood per time per skin area) is at least 10 times $P_m/R_{b/w}$, where $R_{b/w}$, the partition coefficient between blood and water, can be initially estimated as about 1 (Bunge and McCarley, 1996). Various values for cutaneous blood flow are reported in the literature. However, if blood flow is 1.7 mL/hr/cm² (Wade et al. 1962), then blood flow changes will contribute only when $P_m/R_{b/w}$ > about 0.2 cm/hr. However, P_m values exceeding 0.2 cm/hr are usually observed for chemicals with relatively large octanol–water partitioning (i.e., $\log_{10} K_{o/w}$ >~ 3). Because the water solubility limits for such chemicals are extremely low, these chemicals will be seen infrequently in drinking water, and, if present, mass-transfer limitations in the water are likely to be more restrictive than the capacity of the cutaneous blood for removing the penetrating chemical from the skin. Consequently, blood flow resistances are unlikely to be important for chemicals of concern during exposure to drinking water, unless cutaneous blood flow is significantly reduced as might occur by exposure to water colder than skin temperature.

6.2.6 EFFECT OF HYDRATION

The *sc* normally contains 5 to 15% water under normal conditions, but can contain up to 50% when hydrated (Blank and Scheuplein 1964). At normal conditions there is a water concentration gradient across the skin. The innermost layers of the *sc* are in equilibrium with the body tissues, and the outermost layers of the *sc* reflect the humidity of the environment (Blank et al. 1984, Kalia et al. 1996a).

The level of hydration can affect the permeability of the skin to chemicals. Idson (1971) claims that hydration is possibly the most important factor in skin penetration and that increasing hydration increases the absorption of all substances that penetrate the skin. Unfortunately, most studies of the effect of hydration were conducted over hours or days (Behl et al. 1980, Blank et al. 1984, Lambert et al. 1989) and are difficult to relate to bathing or showering exposures of 10 to 30 minutes. Liron et al. (1994) demonstrated that the hydration of dried porcine *sc* exposed to humid air is a slow process, requiring about 4 to 5 hours to reach equilibrium. The skin would be expected to become hydrated more quickly with immersion, but there is no reliable information about the effect of bathing and showering for 10 to 15 minutes on the hydration of the skin.

Chang and Riviere (1991) showed that hydration had no effect on the absorption of parathion in the isolated, perfused porcine skin flap until around 2 hours. Despite permeability enhancement from hydration, the flux of solvents, such as benzene, toluene, and aliphatic alcohols, is less from a saturated aqueous solution than from the neat liquid (Scheuplein and Blank 1973, Dutkiewicz and Tyras 1968, Blank and McAuliffe 1985). In this case, the neat liquids may be better permeability enhancers than water, and absorption from neat liquids should not be used to reflect dermal absorption rates from water.

6.2.7 EFFECT OF SOAPS AND SURFACTANTS

Soaps and surfactants present in showering and bathing scenarios may alter the absorption of chemicals from contaminated water. Strong surfactants such as sodium lauryl sulphate have been shown to enhance the penetration of chemicals (Lindberg et al. 1989) and cause irritation that may subsequently affect penetration (Van der

Valk, Nater, and Bleumink, 1984; Wilhelm, Surber, and Maibach, 1991). One study showed that the absorption of tritiated water and iodine[131] was enhanced three to four times by prewashing rats with a bar of ordinary soap. Soaps may also remove lipids from the surface of the skin and thereby alter penetration characteristics. There is not enough information available to quantitatively estimate the impact of soaps and surfactants on absorption of chemicals from contaminated water.

6.2.8   EFFECT OF IONIZATION

Some chemicals found in water are always ions (e.g., paraquat or tetraethylammonium bromide). Other chemicals (weak acids or bases) can exist in either their ionized or unionized form depending on their pH. Still other chemicals are never unionized, but may be uncharged (i.e., zwitterionic or net neutral) at some pH values. For chemicals with one dominant acid-base reaction, the fraction of chemical that is unionized (i.e., f_{ui}) can be determined from the acid dissociation constant (i.e., pK_a) and the pH of the water in which the chemical is dissolved:

$$f_{ui} = \frac{1}{1 + 10^g} \tag{6.9}$$

where $g = (pH - pK_a)$ for acids and $g = (pK_a - pH)$ for bases. Theoretically, the pH and pK_a values used should be at the temperature of the water to which skin is exposed. However, pK_a values, typically reported at 25°C, are almost never known precisely enough for slight differences in temperature to affect them. As a general rule, an acid is almost entirely unionized if $pK_a - pH > 2$; it is essentially completely ionized if $pH - pK_a > 2$. The situation is reversed for weak bases. For chemicals with potential for multiple ionizations, pK_a values are required for all ionization reactions and mass balances including all possible ionized and unionized species must be conducted (see Vecchia 1997).

The ionic state of a chemical profoundly affects both the rate and extent of its dermal absorption. Generally, the permeability coefficient for a chemical that is almost completely ionized is about two orders of magnitude smaller than when it is almost entirely unionized (Roy and Flynn 1990). (This difference may be slightly smaller for chemicals that are highly hydrophilic even in the unionized form.) To a good approximation, penetration can be attributed to the unionized species alone, particularly when the unionized fraction is not too small (i.e., $f_{ui} >$ about 0.1). Many studies have found that absorption rates for ionizing organic chemicals are accurately correlated with the unionized fraction (e.g., Roy and Flynn 1990, Parry et al. 1990). That is, dermal absorption rates are correctly estimated by multiplying the concentration of the unionized species by the permeability coefficient measured at a pH when the chemical was completely unionized.

With few exceptions, chemicals that are entirely ionized will penetrate skin too slowly to be of systemic health concern (except for chemicals present at high concentrations or highly toxic chemicals). However, completely ionized chemicals may still have localized effects in the form of irritation or allergic reactions of the skin itself.

6.2.9 Consideration of Special Populations

Differences in skin barrier function for special populations such as pregnant women or children should be no greater than the variations within the general public. Thus estimates of skin parameters for the general public should still apply to these groups. An exception to this conclusion, premature, low-birth-weight infants have a poorly developed stratum corneum (Kalia et al. 1996) and are at greater risk from dermal exposure. However, recent measurements have shown that these infants developed a competent barrier within 4 weeks postnatal age, regardless of their gestational age at birth (Kalia et al. 1996).

6.2.10 Summary

Although many factors will affect dermal absorption from household drinking water exposures (primarily showering and bathing), a number of simplifications are reasonable for preliminary estimates. The *sc* will present the rate-limiting barrier in most cases. The ionic condition of the chemical significantly affects the extent and rate of dermal absorption and should be considered. The effects of regional variations, elevated temperature, hydration, soaps, and surfactants are difficult to quantify and probably contribute modestly in most cases. The permeability coefficient quantitatively defines the ability of a chemical to penetrate through skin. Steady-state permeability coefficients are required for comparing one chemical with another and for developing correlations of existing permeability data.

6.3 TYPES OF SKIN PENETRATION DATA

Dermal penetration is measured and reported in several ways. Measurements include the total amount that penetrates the outer surface of the skin, the amount that is present in the skin, the amount that has permeated across the entire skin layer and entered the systemic circulation or the receiving chamber of a diffusion cell, and the rate at which the chemical enters or leaves the skin. These measurements are discussed next.

6.3.1 Ways to Express Penetration

6.3.1.1 Percent Absorbed

Several investigators have reported dermal absorption as the percent of the exposed dose that absorbs. Unfortunately, percent absorbed values from the laboratory can only be used to estimate human absorption when all exposure conditions are identical (e.g., exposure time, volume of solution, the animal species, and in some cases exposure concentration). If the chemical is not lost from the surface during the exposure, the percent absorbed will be small for short exposure times (approximately 0% for an exposure time of zero), and almost 100% for very long exposure times. If the percent absorbed is less than 10 to 20% of the exposed dose, increasing the volume of the exposed dose without changing the concentration will produce almost no change in the amount absorbed, causing the percent absorbed to decrease (because the exposed dose increased). That is, percent-absorbed values are not independent of the mass of the exposed dose, which has been supported by experimental results (Scheuplein and Ross 1974, among others). As a result, experimental percent absorbed values are not readily extractable to actual exposure scenarios, particularly for aqueous solutions, in which the volume of water (and consequently the exposed dose) and the exposure time can vary significantly from one exposure situation to another.

6.3.1.2 Flux

Flux, or mass per surface area exposed per time (i.e., mg/cm^2/s), is a common way of expressing results from dermal absorption experiments. When a large amount of chemical is placed on the surface of the skin, *in vitro* or *in vivo*, flux across the skin increases during the exposure until a steady-state value is reached (see slope of the $\tilde{M}_{in}$ curve shown in Figure 6.2). Most experimental measurements of flux require that the concentration on the skin surface and the surface area exposed are constant. At steady-state, the mass absorbed per time becomes constant. Flux from *in vitro* diffusion cell studies is calculated using the slope of the linear part of the accumulated amount of chemical that has penetrated over time plot (see $\tilde{M}_{out}$ curve in Figure 6.2). To calculate flux from *in vivo* experiments, it is necessary to estimate the total amount of chemical absorbed during the exposure. The whole animal complicates the analysis, because direct measurement is difficult, as described in the *in vivo* experimental methods below. The flux measured depends on the exposure conditions; flux varies with concentration as well as time. Measurements of flux are most useful when the laboratory situation mimics the actual situation of interest. Flux is the preferred way to express absorption rates from exposure to neat liquids. Flux is always more useful than percent absorbed because it can be used to extrapolate absorbed dose for other surface areas and times.

6.3.1.3 Permeability Coefficients

The permeability coefficient (expressed in units of distance per time) is a common way to express results from dermal absorption experiments. Determination of permeability coefficients *in vitro* and *in vivo* generally requires that the exposure concentration and surface area are known and constant. Permeability coefficients are most commonly estimated by dividing the steady-state flux from diffusion cells or whole animal studies by the exposure concentration. It is also possible to use physiologically based pharmacokinetic models to estimate the permeability coefficient, as will be discussed below.

Permeability coefficients are time independent provided the vehicle and chemical do not alter the skin barrier, and that the reported permeability coefficient was calculated after the dermal absorption rate had reached steady-state. Permeability coefficients are generally assumed to be concentration independent and are therefore useful for extrapolating from one exposure concentration to another.

Permeability coefficients vary depending on the vehicle (if any) used in the experiment. This is because the permeability coefficient is proportional to the equilibrium partition coefficient between the *sc* and the vehicle (defined as $R_{sc/v}$), which will be different for different vehicles (i.e., for nonaqueous vehicles, $R_{sc/v}$ replaces

$R_{sc/w}$ in Equation (6.6)). This concept is very important because the permeability coefficients for benzene penetration from a saturated vapor, for a pure liquid, and from a saturated aqueous solution would all be different even if the flux from each medium were similar. Appropriate permeability coefficients are the most useful way to express penetration from chemical solutions such as contaminated water.

6.3.1.4   Partition Coefficients

Although partition coefficients between the skin ($R_{m/w}$) (or the individual skin layers, $R_{sc/w}$ and $R_{ve/w}$) and water do not indicate penetration, they do indicate the capacity of the skin to absorb a chemical. Also, the penetrability of the skin depends on the ability of the permeating chemical to absorb into the skin. To be consistent with the definition of permeability as described by Equations (6.2), (6.4), and (6.6), the partition coefficient is calculated as the equilibrium concentration in the skin layer (mass of chemical per volume of the specified skin layer at absorption conditions) divided by the concentration in solution (mass of chemical per volume of solution) (Parry et al. 1990). Often, whole or split-thickness skin is used in partition coefficient measurements. The partition coefficient for a skin sample containing both sc and ve ($R_{m/w}$) will depend on the partition coefficients for the individual layers as:

$$R_{m/w} = \frac{L_{sc}}{L_m} R_{sc/w} + \frac{L_m - L_{sc}}{L_m} R_{ve/w} = R_{sc/w}\left[\frac{L_{sc}}{L_m} + \frac{L_m - L_{sc}}{L_m} R_{ve/sc}\right] \qquad (6.10)$$

If the absorbing chemical is lipophilic, then $R_{ve/sc}$ is often small and consequently the sc partition coefficient is equal to the partition coefficient for the entire skin sample divided by the sc fraction of the total skin sample.

Partition coefficient values are ratios of concentration in the sc and water and unitless, when the same concentration units are used for both the sc and water. Partition coefficients calculated on the basis of skin mass and solution mass (i.e., the mass of absorbing chemical in the skin per unit mass of dry skin divided by the equilibrium mass of chemical in the solution per unit mass of solution) will be about 10 to 100 times larger than values calculated based on the skin volume at absorbing conditions (see Parry et al., 1990 and Vecchia, 1997). Since both mass and volume based partition coefficients are unitless, it is difficult to know which type is reported, unless the authors have specified (which does not always happen).

6.3.1.5   Lag Time

Lag time is seldom used alone to describe results of skin absorption experiments, but $t_{lag,m}$ can provide useful information. Lag time is an indication of the time it takes for a chemical to penetrate through the skin. If the sc is undamaged, $t_{lag,m}$ is nearly equal to the lag time for the sc alone (i.e., $t_{lag,sc}$), unless P_{sc}/P_{ve} > about 0.1 (see Equation (6.7)). It is directly related to the skin thickness and inversely related to diffusivity (see Equations (6.3) and (6.8)). It is often determined experimentally by extrapolating the slope of the linear part of the accumulated amount absorbed over time plot back to the time axis (see Figure 6.2). However, lag times measured in this way are notoriously variable and occasionally produce physically impossible results (e.g., negative lag times are indicated). Among other factors, minor variations in the slope of the apparent steady-state data can dramatically affect the time-intercept.

6.3.2 EXPERIMENTAL METHODS

Most skin penetration measurements fall into one of two categories: *in vivo* experiments made in humans or living animals, and *in vitro* experiments made in diffusion cells with excised skin from humans or animals. There is controversy about whether *in vitro* or *in vivo* measurements are the most appropriate ways to measure absorption of chemicals. Researchers tend to provide evidence that their own method is most representative of human skin penetration. *In vitro* methods can provide quick and direct measures of flux and permeability coefficients. They also have the advantage that human skin can be used when the chemical would be too toxic for *in vivo* human studies. Although the actual *in vitro* experiment is simpler, it requires many important decisions if the data are to be representative of the human. These choices include animal species, thickness of skin, source (fresh or frozen) of skin, and receptor solution. *In vivo* studies come with their own problems and choices. *In vivo* studies are often more elaborate and require more data analysis. The preferred method for risk assessment purposes, when *in vivo* human studies are not possible, varies according to how the data will be used.

6.3.2.1 *In Vitro*

Skin permeability is frequently measured in a diffusion cell where skin is mounted between two well-stirred solution chambers. Typically, the aqueous donor chamber contains a known mass of solute, and its concentration in the receiving chamber is monitored. The cumulative mass of solute appearing in the receiving chamber is then plotted as a function of time since the exposure began. Provided that the vehicle concentration remains essentially constant and that sink conditions are maintained in the receiving chamber, the cumulative mass of solute appearing in the receiving chamber eventually becomes linear in time, indicating that steady-state has been established. The permeability coefficient is calculated from the slope of the linear regression of the cumulative mass vs. time (i.e., the slope = $P_m C_v^o A$). The intercept of the extrapolation of the steady-state line to a cumulative mass of zero is defined as the lag time ($t_{lag,m}$). A theoretical description of an *in vitro* experiment (assuming skin behaves as a single homogeneous membrane) is shown in Figure 6.2. In this plot, the cumulative mass plotted on the ordinate is normalized by concentration and area ($R_{m/w} C_w^o L_m A$) and exposure time has been normalized by the characteristic diffusion time (L_m^2/D_m).

Several situations can compromise the accuracy of a permeability coefficient determined in this way. A slope analysis as described above requires that the donor chamber concentration not decrease by more than about 10 to 20%, and that sink conditions be maintained in the receiving chamber. If the donor chamber concentration depletes or the receiving chamber concentration builds, then the simple slope

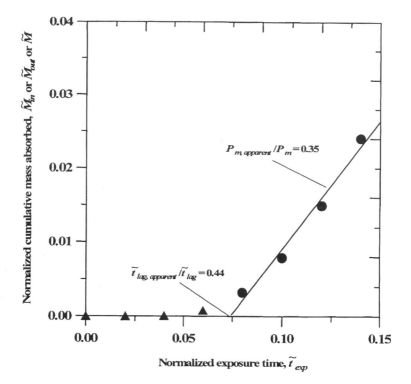

FIGURE 6.3   Illustration of the uncertainties in graphically determining t_{lag}.

analysis will underestimate the actual permeability; the permeability can be determined in this situation using a more complicated data analysis (Cussler 1984). More commonly, the diffusion chamber and experimental procedure are designed to provide constant donor concentration and sink conditions, often by periodically replacing the receptor with a flow-through arrangement.

The permeability coefficient is also underestimated if any of the data used in the slope analysis were obtained under conditions that were not at steady-state. Confirming steady-state conditions is not as simple as it seems. The standard approach of using those data that appear to be linear is extremely sensitive to the frequency and range of the data collected. Obviously, data from no fewer than four different exposure times are required to indicate linearity. Even then, non-steady-state data may appear to be linear if the data range is too narrow. For example, in Figure 6.3, data that fit perfectly on the $\tilde{M}_{out}$ curve from the non-steady-state Region A (shown as *data points*) in Figure 6.2 look linear, even though the more complete data set (i.e., the $\tilde{M}_{out}$ curve in Figure 6.2) shows that the data do not vary linearly with time. A linear analysis of the data in Figure 6.3 produced an apparent permeability coefficient and an apparent dimensionless lag time, both of which are lower than the actual values (e.g., the apparent P_m = 0.35 actual P_m and the dimensionless apparent lag time of 0.075 is 0.44 of the actual lag time). Based on theoretical considerations for the length of time required to reach steady-state, Bunge et al. (1995) have proposed requiring all data used in the linear slope analysis must be for times $> 2.4 \, t_{lag,m}$. Data points from exposure times less than this may not be at steady-state. Using this guideline, the data points used in the linear analysis illustrated in Figure 6.3 should have been from normalized exposure times > 0.18 ($=$ 2.4 * 0.075). Thus, this guideline indicates (as it should) that the data used in the linear analysis in Figure 6.3 were not at steady-state.

Most of the errors associated with collecting *in vitro* permeability data in this way lead to underestimates of the actual permeability coefficient. As discussed shortly, most of the errors associated with collecting and analyzing *in vivo* data cause the experimental permeability coefficient to be larger than the actual value. Therefore, the widely stated observation that *in vivo* permeability coefficients are larger than those measured *in vitro* is not surprising and may not reflect an actual difference between *in vivo* and *in vitro* values, but errors in the procedure and analysis.

Even with a flow-through arrangement for the receptor fluid, providing sink conditions for highly lipophilic chemicals may be difficult. Some studies have tried to improve solubility of the receptor fluid with additives such as albumin, surfactants, or water-soluble organic solvents such as ethanol. These measurements often show an apparently larger permeability when additives were present. Many additives, however, are known or suspected to be damaging to the skin, making it difficult to know whether the larger permeability measured in their presence is real or an artifact of an altered membrane. Several investigators simultaneously monitor permeability of the skin sample to tritiated water, as an indicator of damage, discarding solute permeability data whenever the tritiated water permeability exceeds an acceptable value.

A few variations of this *in vitro* diffusion cell experiment are common. Occasionally, chemical is applied to the skin surface on the donor side using a volatile solvent, such as acetone or methanol. The solvent evaporates within a few minutes leaving the chemical deposit, usually with a planned surface loading of 1 to 10 $\mu g/cm^2$. These are called finite dose experiments, because dermal absorption usually depletes a significant percentage of the donor mass. Permeability coefficients are difficult to estimate from data in this form, and data from this type of experiment are commonly reported as the percent of the applied dose that penetrates. As indicated by Equation (6.2), the permeability coefficient for chemical transfer across a membrane from water depends on the equilibrium partition coefficient of the chemical between the membrane and water (i.e., $R_{m/w}$). When a chemical is presented to the skin from a media (a vehicle, v) other than water, the permeability coefficient will depend on the partition coefficient between the membrane and the vehicle (i.e., $R_{m/v}$) instead of $R_{m/w}$. Consequently, permeability coefficients for a given chemical are different for different vehicles. Thus it is inappropriate to use unadjusted permeability coefficients measured for neat chemicals or from solutions of nonaqueous vehicles to indicate the permeability from water.

A few investigators have measured the amount of solute in the skin in addition to that appearing in the receiving chamber. If steady-state has been reached, the mass of solute in the skin reflects the skin's affinity for the solute and is related to the partition coefficient between the *sc* and the vehicle (usually water). For some highly lipophilic chemicals, almost all of the solute will be in the skin with little in the receiving chamber. Under these circumstances, the required data analysis must consider the mass of chemical absorbed into skin in addition to the mass of chemical penetrated through the skin.

6.3.2.2   *In Vivo*

In vivo studies in laboratory animals may be more representative of human exposures than *in vitro* exposures because of intact blood flow, immune responses, and metabolism; however, determination of absorption rates is a much more complicated process because the rest of the animal is attached to the skin. To perform a good *in vivo* study, the researcher must accurately determine the total amount of chemical that has been absorbed, and precisely control the area of skin exposed. Occlusion of the site of application complicates some *in vivo* studies. The chemical on the surface is often covered by an impermeable dressing so that loss of the chemical (from evaporation or being rubbed off) is minimized during the exposure. Occluding the site of application inhibits normal transepidermal water loss and causes the skin under the dressing or patch to become more hydrated and may impact the rate of penetration (see discussion about the effect of hydration above).

There are two general approaches to *in vivo* penetration studies, indirect (non-invasive) measurements and direct measurements. Indirect measurements can be obtained by simply applying a known amount of chemical, allowing it to remain in contact with the skin for the experimental period, and then estimating how much chemical was absorbed by determining the difference between the mass of chemical remaining on the skin and the mass applied. This approach has been used successfully for chemicals that are not volatile. Other indirect methods measure the elimination of the absorbed chemical from the animal. In these types of studies, the total mass absorbed is estimated by measuring urinary and fecal excretion over a long period of time after the experiment. The assumption is made that total absorption equals the total amount of chemical eliminated. Sometimes a correction factor for the mass of chemical that remains in the animal (or is otherwise lost) is made by dosing the animal intravenously and determining the percentage of chemical eliminated in urine and feces. Elimination can be nonlinear and therefore it is important to match the intravenous amount to the estimated total amount absorbed for this correction factor to be reasonably accurate. Indirect measurements can be simple and useful in some situations. They generally provide one average measure of absorption during the exposure period. As such, variations in absorption rates with exposure time are lumped together. Since absorption rates are faster at the initiation of the experiment than after steady-state is reached, the absorption rates reported from experiments of this type are probably larger than steady-state absorption rates.

Direct measurements of a chemical in skin, blood, and tissues are another, more complicated *in vivo* approach. The challenge with direct measurements is estimating the total absorbed in the whole body based on the amount measured in blood or tissue. Pharmacokinetic models (which will be described in detail later) may be used to estimate total absorption from blood, exhaled breath, or tissue concentrations in direct studies. Another possible approach is to measure the amount in the whole animal by measuring every tissue and fluid. Elimination cannot be ignored in direct studies, and therefore must also be accounted for in estimating total mass absorbed. Direct studies using blood or exhaled breath measurements can employ serial measurements during and after the exposure and therefore can provide more detailed information about the time course of absorption than indirect measurements. Direct measurements in skin or other organs require serial sacrifices in order to obtain information on the time course of absorption.

Another commonly used method for direct measurement of chemical in the skin is known as "tape-stripping." With this technique, adhesive tape is applied firmly to the skin and repeatedly pulled off. Each repetition removes the outer cells of the *sc* that adhere to the tape. Each tape-strip removes successively fewer cells until the remaining cells are firmly attached. Because the mass of *sc* removed with each tape-strip varies, it is usually necessary to weigh each strip to determine the mass of *sc*. The number of tape-strips required to remove the entire *sc* is variously reported to be 20 to 40 and depends on the type of adhesive tape used, the skin site, and the individual. Tape-stripping can be useful for determining the mass of chemical in the outermost part of the *sc*. However, to use tape-strip data for estimating permeability coefficients requires data collection immediately after exposure (i.e., during the non-steady-state period; Pirot et al. 1997). Complications of this method are that (1) the chemical continues to diffuse across the *sc* while tape-stripping occurs, requiring the stripping procedure to be rapid, (2) the quantity of chemical on the tapes can be quite small, making reliable chemical analysis difficult, and (3) data variability tends to be quite large.

6.3.3 Sources of Penetration Data

Dermal penetration of chemicals from aqueous solutions have been studied by many investigators for a large number of chemicals. These measurements are distributed through many research publications. A few authors have compiled lists of some of these measurements. Collections of *in vitro* permeability coefficients are most common (e.g., Brown and Rossi 1989, Kasting et al. 1992, McKone and Howd 1992, Potts and Guy 1993, to name a few data collections listing data primarily from excised human skin). Flynn (1990) assembled 97 human skin (mostly *in vitro*) permeability coefficients for 94 compounds with a relatively broad range of properties ($18 < MW < 765$ and $-3 < \log_{10}K_{ow} < 6$). The Flynn database was included in the U.S. EPA document *Dermal Exposure Assessment: Principles and Application* (U.S. EPA 1992). While much penetration data can be found in the literature, collections of *in vitro* permeability coefficient data from animal skins and skin–water partition coefficient data from either human or animal skin are much more limited. As discussed earlier, *in vivo* absorption data usually require pharmacokinetic analysis to extract parameters such as the permeability coefficient, which can then be compared to *in vivo* or *in vitro* measurements of other chemicals. In other cases, indirect *in vivo* measurements make direct comparisons difficult to interpret. As a consequence, only a few collections with limited *in vivo* data exist, although the total number of experiments distributed through the dermal absorption literature is reasonably large.

Data quality is clearly important yet difficult to judge, perhaps more so for *in vivo* experiments, in which the chemical and data analyses are more complex. Some minimum quality criteria are reasonable requirements (e.g., demonstrated steady-state and constant vehicle concentration for *in vitro* permeability coefficient data). Vecchia (1997) has attempted to assemble and qualify the published human and animal skin data for *in vitro* steady-state permeability coefficients from water and *in vitro* equilibrium skin–water partition coefficients. The resulting databases (1) are significantly larger than previous databases, (2) distinguish meaningful measurements from those that cannot be interpreted, and (3) contain relevant information (e.g., experimental temperature, and values for pK_a, and $K_{o/w}$) that was not originally reported (either obtained through communication with the authors or calculated).

6.3.4 CORRELATIONS FROM MANY DATA SETS

When a large body of data are available (as is the case for *in vitro* permeability coefficients), one valid approach for extending the data to chemicals that have not been measured is to derive equations that correlate the observations. Such correlations can be empirical or theoretical. The empirical approach accepts any equation form that fits the data. Theoretical approaches choose a general equation that is consistent with the physicochemical mechanisms that are thought to operate. To the extent that the hypothesized mechanism represents reality, one can extrapolate to a chemical that is not included in the database more confidently from a theoretical correlation than from an empirical correlation.

This leads to the question: Which is more useful: the correlation-predicted value, or, if it is available, an experimental value? There is no easy answer. Many risk assessors will argue to "always use appropriate data, if available." However, the facts are that all data involve variability, which is often unquantified unless repeated experiments were conducted, and some data are obtained using less than optimal procedures. Examples of this second situation are permeability coefficient values calculated from data in which the receptor chamber solution was not at sink conditions or in which the skin was damaged. Theoretically based correlations will average the experimental variability for the entire database. Consequently, when differences arise between the correlation and experimental values of parameters like the permeability coefficient, the weight of the experimental data (that is, the large number of experiments included in the database on which the correlation is based) strongly support the validity of the correlation prediction over that single experimental data point. That is, statistically, chances are greater that the correlation represents truth compared to a value from a single experiment. Of course, agreement between repeated measurements, especially by more than one laboratory, statistically support the experimental values over the correlation. Significantly, Vecchia (1997) found that *in vitro* permeability coefficients reported by different laboratories for the same chemical often differ by at least one order of magnitude.

Many investigators have proposed different correlations for predicting permeability coefficients for chemicals penetrating human skin from a water vehicle. Several of these correlations are functions of the octanol–water partition coefficient ($K_{o/w}$) and/or molecular weight (*MW*). A long, but still incomplete list, is shown in Table 6.2, which also lists the chemical class(es) and the range of $K_{o/w}$ and *MW* (noted as suggested range) of the data on which each correlation was based. Figures 6.4 through 6.6 compare the correlations described in Table 6.2 with experimental data (from the Flynn permeability database; Flynn 1990) for chemicals with $MW \approx 100$, 300 and 500.

Based on the results shown in Figures 6.4 through 6.6, several observations can be made. First, many of the correlations predict a similar dependence on $K_{o/w}$. The distinctive differences between correlations are in their prediction of the effect of MW. Second, nearly all of the correlations reasonably predict the experimental data for chemicals of $MW \approx 100$. (Exceptions are models numbered 2, 6, 7, and 16.) Third, the differences between correlations and between predicted and experimental values increase with MW of the penetrating chemical. Finally, some of the correlations in Table 6.1 were developed from a data set with a limited range of properties, and, not surprisingly, those correlations do not perform well when used outside of that range. Examples are (1) model numbers 11 and 14, which were developed for chemicals with lower MW (and therefore perform well at lower MW and more poorly at higher MW); and (2) model number 16, which was developed for higher MW chemicals (and therefore performs well at higher MW but more poorly at lower MW). The clear message is that correlations should not be used outside of the range of properties of the data set from which they were developed. As shown in Figures 6.4 through 6.6, the Potts and Guy correlation (Model number 15 in Table 6.2):

$$\log_{10} P_m (\text{cm/hr}) = -2.72 + 0.71 \log_{10} K_{o/w} - 0.0061 \, MW \qquad (6.11)$$

is among the better correlations for representing experimental permeability data for a broad range of chemicals. Despite the adequate correlations, as shown in Figure 6.7 (from Vecchia, 1997), experimental data may deviate from predictions made using Equation (6.11) by an order of magnitude. Differences in experimental temperature are partly responsible for this variability. However, data uncertainty is a more likely cause of significant difference between predictions and experiments. For example, Vechia (1997) showed that measurements replicated for the same chemical in different laboratories often vary by an order of magnitude or more.

Equation (6.11) was used by EPA to estimate permeability coefficients for a list of about 200 pollutants (U.S. EPA 1992). Because the database on which Equation (6.11) is based (i.e., from Flynn 1990) consists largely of unionized chemicals, Equation (6.11) should represent the permeability for unionized organic chemicals. Although the Flynn (1990) database upon which Equation (6.11) is based does include highly lipophilic chemicals, the data suggest that $P_{sc} \cong P_m$, probably because the MW of these chemicals is so large that P_{sc} is still less than P_{ve}.

Correlations like those listed in Table 6.2 assume that MW is a good predictor for molecular size. This assumption may not be appropriate for groups of compounds with chemical diversities affecting molecular size. For example, chlorinated hydrocarbons will occupy the same molar volume as a hydrocarbon molecule with a much smaller MW. As a result, MW correlations developed from databases of primarily hydrocarbons will tend to underestimate permeability coefficients for chemically dense compounds such as chlorinated hydrocarbons (typically by about a two- to fourfold).

Correlations also exist for skin–water partition coefficients, although fewer than for permeability coefficients. Like equations to predict permeability coefficients, equations to predict partition coefficients should not be used outside the range of

TABLE 6.2
Permeability Coefficient Correlations Based on $K_{o/w}$ and MW

Model	Equation Reference	Chemical Class[a]	Permeability Correlation (P_m in cm/hr)	Data Reference	Data Range
1	Abraham et al. (1995)	Misc (n = 43)	$\log P_m = -2.184 + 0.851 \log K_{o/w} - 0.012 MW$	Not specified[b]	Not specified[b]
2	Vecchia (1997)	Misc (n = 170)	$\log P_m = -2.44 + 0.514 \log K_{o/w} - 0.0050 MW$	Vecchia (1997)	$18 < MW < 585$ $-3.1 < \log K_{o/w} < 4.6$
3[c]	Bronaugh & Barton (1991)	Flynn Database (n=90)	$\log P_m = -2.61 + 0.67 \log K_{o/w} - 0.0061 MW$	Flynn (1990)	$18 < MW < 765$ $-3 < \log K_{o/w} < 5.5$
4	Brown & Rossi (1989)	Alkanols, phenols, drugs (n = 39)[d]	$\log P_m = -1.0 + 0.75 \log K_{o/w} - \log\left(120 + K_{o/w}^{0.75}\right)$	Scheuplein (1973), Roberts (1977), Michaels (1975)	$32 < MW < 765$ $-0.8 < \log K_{o/w} < 4.6$
5	Cleek & Bunge (1993)[e]	Flynn Database (n=90)	$\log P_{sc} = -2.8 + 0.74 \log K_{o/w} - 0.006 MW$ $$P_m = \frac{P_{sc}}{1 + \dfrac{P_{sc}\sqrt{MW}}{2.6}}$$	Flynn (1990)	$18 < MW < 765$ $-3 < \log K_{o/w} < 6$
6	El Tayar et al. (1991)	Steroids (n = 11)	$\log P_m = -5.324 + 0.8 \log K_{o/w}$	Anderson et al. (1988)	$403 < MW < 503$ $1.4 < \log K_{o/w} < 5.5$

	Reference	Database	Equation	Ref.	Range
7	El Tayar et al. (1991)	Phenolics (n = 18)	$\log P_m = -5.5154 + 2.39 \log K_{o/w} - 0.37\left(\log K_{o/w}\right)^2$	Roberts et al. (1977)	$94 < MW < 197$ $0.8 < \log K_{o/w} < 3.7$
8	Flynn & Amidon (in US EPA, 1992)	Flynn Database (n=90)	$\ln P_m = -3.31 + 0.79 \ln K_{o/w} = -1.45 \ln MW$	Flynn (1990)	$18 < MW < 765$ $-3 < \log K_{o/w} < 6$
9	Flynn (1990)	Flynn Database (n=90)	For MW ≤ (150) $\log P_m = -3.0 \;\; \left(\log K_{o/w} < 0.5\right)$ $\log P_m = -3.5 + \log K_{o/w} \;\; \left(0.5 \le \log K_{o/w} \le 3.0\right)$ $\log P_m = -0.5 \;\; \left(\log K_{o/w} > 3.0\right)$ For MW > 150 $\log P_m = -5.0 \;\; \left(\log K_{o/w} \le 0.5\right)$ $\log P_m = -5.5 + \log K_{o/w} \;\; \left(0.5 \le \log K_{o/w} \le 3.5\right)$ $\log P_m = -1.5 \;\; \left(\log K_{o/w} > 3.5\right)$	Flynn (1990)	$18 < MW < 765$ $-3 < \log K_{o/w} < 6$
10	Kasting et al. (1992)	Misc.[f]	$P_m = \left[\dfrac{1}{P_{lip} + P_{pol}} + \dfrac{1}{P_{aq}}\right]^{-1}$ $\log P_{lip} = -2.87 + \log K_{o/w} - 0.0078 MW$ $P_{pol} = 1\times10^{-5}\sqrt{300/MW}$ $P_{aq} = 0.15\sqrt{300/MW}$	Detailed in Ref.	$18 < MW < 518$ $-1.4 < \log K_{o/w} < 6.3$

TABLE 6.2 (continued)
Permeability Coefficient Correlations Based on $K_{o/w}$ and MW

Model	Equation Reference	Chemical Class[a]	Permeability Correlation (P_m in cm/hr)	Data Reference	Data Range
11	McKone & Howd (1992)	Misc.[g] (n = 51)	$\log P_m = 2.602 + \log\left(2.4\times10^{-6} + 3\times10^{-5} K_{o/w}^{0.8}\right) - 0.61 \log MW$	Detailed in Ref.	$18 < MW < 227$ $-1.4 < \log K_{o/w} < 4.6$
12	Michaels et al. (1975)[h]	Drugs (n = 10)	$\log P_m = -3.6 + \log K_{m/w} + \log\left(1.16 + 3\times10^{-6} K_{m/w}\right)$ $\quad - \log\left(0.16 + 2\times10^{-3} K_{m/w}\right)$ $K_{m/w}$ = mineral oil / water partition coefficient $\left(\text{Plotted results assumed } K_{m/w} \approx K_{o/w}\right)$	Michaels et al. (1975)	$165 < MW < 765$ $1 < \log K_{o/w} < 4$
13	Morimoto et al. (1992)	Drugs (n = 16)	$P_m = 4.21\times10^{-4} K_{o/w}^{0.75} + 9.83\times10^{-5}$	Morimoto et al. (1992)	$130 < MW < 358$ $-4.7 < \log K_{o/w} < 4.0$
14	Potts & Guy (1993)	n-Alkanols, acids, diols (n = 23)	$\log P_m = -2.24 + 0.81 \log K_{o/w} - 0.013 MW$	Scheuplein et al. (1967)	$18 < MW < 160$ $-1.4 < \log K_{o/w} < 3.0$
15	Potts & Guy (1993)	Flynn database (n≈90)	$\log P_m = -2.72 + 0.71 \log K_{o/w} - 0.0061 MW$	Flynn (1990)	$18 < MW < 765$ $-3 < \log K_{o/w} < 6$
16	Siddiqui et al. (1989)	Steroids (n = 7)	$\log P_m = -6.66 + 1.05 \log K_{o/w}$	Siddiqui et al. (1989)	$288 < MW < 476$ $1.0 < \log K_{o/w} < 3.5$

Source: Brent E. Vecchia, M.S. Thesis, Colorado School of Mines, 1997 (see reference).

a General description of chemicals; n = the total number of data points; the number of different chemicals may be fewer.

b Although not directly specified, there is some indication that this data is largely composed of the alkanols, alkanoic acids, and alkanediols of Scheuplein and Blank (1967) and the phenols reported by Roberts et al. (1977).

c The Bronaugh and Barton model (#3) is essentially identical to the Potts and Guy model (#15).

d Data adjusted to 31°C by assuming the permeability coefficient doubles for every 10°C increase in temperature.

e Modification of Potts and Guy model (#15).

f This database of more than 130 permeability measurements shares some permeability measurements with the Flynn database. Most of the additional values were measured either with shed snake skin as the membrane (13 values), or with human skin from the vehicle propylene glycol (78 values).

g Includes many of the permeability measurements for chemicals of $MW < 227$ tabulated in the Flynn database. In vivo guinea pig permeability measurements for three chemicals were also included.

h Lipophilicity is represented with the mineral oil–water partition coefficient rather than the more coventional log K_{ow}. For unknown reasons, these two measures of lipophilicity are poorly correlated for the ten chemicals, but log K_{ow} tend to be higher on average; when logK_{ow} are substituted for the mineral oil–water partition coefficients the permeability coefficient is overestimated.

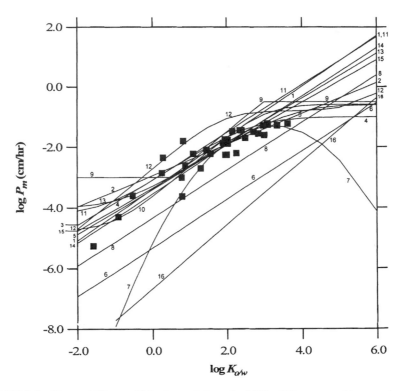

FIGURE 6.4 Permeability coefficient correlations for $MW = 100$ compared to experimental data from the Flynn permeability database (Flynn 1990) for chemicals of $50 < MW < 150$ (from Vecchia 1997).

conditions for which they were developed. Figure 6.8 (from Cleek and Bunge 1993) shows skin–water partition coefficients for steroids (from Scheuplein et al. 1969), phenolic compounds, and aromatic alcohols (from Roberts et al. 1977), and normal alcohols and small undissociated acids (from Scheuplein 1967). The resulting regression equation:

$$logR_{sc/w} = -0.006(\pm0.21) + 0.57(\pm0.04)logK_{o/w} \tag{6.12}$$

with a correlation coefficient of 0.85 is indicated as the solid line and does not include the data for the more hydrophilic chemicals represented by triangles. Despite some data scatter, the highly hydrophilic compounds with log $K_{o/w} < $ -0.28 (i.e., water, methanol, ethanol, formic acid, and acetic acid) appear to reach a constant $R_{sc/w}$ of about 0.9. The dashed line in Figure 6.8 corresponds to the $K_{o/w}$ portion of the permeability correlation proposed by Bunge et al. (1994), a slight modification of the correlation from Potts and Guy (1992), log $R_{sc/w} = 0.74$ log $K_{o/w}$, which to be consistent with Equation (6.11) is more appropriately used as:

$$logR_{sc/w} = 0.71\, logK_{o/w} \tag{6.13}$$

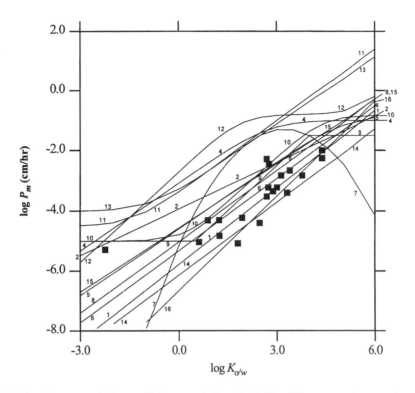

FIGURE 6.5 Permeability coefficient correlations for $MW = 300$ compared to experimental data from the Flynn permeability database (Flynn 1990) for chemicals of $250 < MW < 350$ (from Vecchia 1997).

Although Equation (6.12) clearly provides a better fit, statistically, the difference between Equations (6.12) and (6.13) is not significant.

Equation (6.12) and Equation (6.13) should both represent unionized organic chemicals. For exposures to chemicals that dissociate, the concentration of the unionized species is appropriate. That is, for estimating the amount of dermal absorption, the correct value for C_w^o is:

$$C_w^o = C_{total} \, f_{ui} \tag{6.14}$$

where the unionized fraction is calculated knowing the pH of the solution and the pK_a of the chemical as described in Equation (6.9).

Cleek and Bunge (1993) used a slight variation of the Potts and Guy equation for estimating permeability coefficients (i.e., Model number 15 in Table 6.2) together with Equation (6.13) for the sc–water partition coefficient in a simple approximation of the membrane description for skin. The resulting equations, which are discussed more in the next section, predict absorption rate behavior consistent with that illustrated in Figures 6.1 and 6.2. These equations provide a theoretically consistent approach for using steady-state and equilibrium data (i.e., permeability coefficients

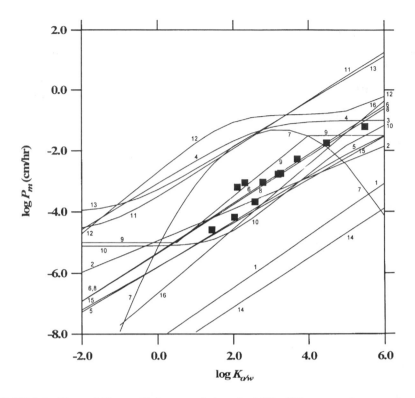

FIGURE 6.6 Permeability coefficient correlations for $MW = 500$ compared to experimental data from the Flynn permeability database (Flynn 1990) for chemicals of $450 < MW < 550$ (from Vecchia 1997).

and partition coefficients) to estimate the more rapid absorption during short exposures such as those that occur in showering or bathing.

Bogen (1995) presented a regression equation developed from *in vivo* estimates of effective (i.e., time dependent) permeability coefficients for nine chemicals measured in humans (aniline, carbon disulfide, chloroform, ethylbenzene, styrene, and toluene) and guinea pigs (2-butoxyethanol), and hairless guinea pigs (chloroform, tetrachloroethylene, and trichloroethylene). Specifically, Bogen (1995) used published *in vivo* uptake data for exposure times of between 10 and 120 minutes to represent effective (i.e., time dependent non-steady-state) permeability coefficients for a 12-minute exposure. He regressed these data as functions of $K_{o/w}$ and MW to produce a correlation similar in form to those listed in Table 6.2. Permeability coefficients estimated from Bogen's correlation should be used with Equation (6.4) and, distinctive from estimates made using equations in Table 6.2 (which are based on steady-state measurements), do not apply to longer time exposures.

Cleek and Bunge (1993) proposed an approach for estimating a time-dependent effective permeability from steady-state (time-independent) values. On average, the 12-minute *in vivo* data estimated by Bogen (1995) were two- to fivefold higher than those calculated by the Cleek and Bunge (1993) approach (see Section 6.4.1). This

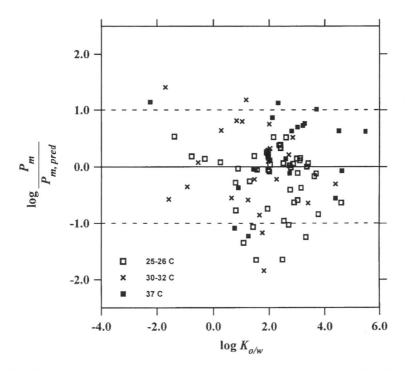

FIGURE 6.7 Comparison of experimental permeability coefficients from the Flynn database (Flynn, 1990) with predictions from Equation (6.11) (from Vecchia, 1997).

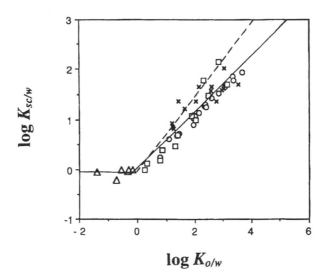

FIGURE 6.8 $K_{sc/w}$ as a function of $K_{o/w}$ for normal alcohols (□), undissociated acids (△), steroids (✘), and phenolic compounds and aromatic alcohols (○) compared to Equation (6.13) (− −) and Equation (6.12) (——) (from Cleek and Bunge, 1993).

difference may not be surprising, because the Cleek and Bunge (1993) approach is based on steady-state permeability measurements on human skin primarily from abdomen, back, or legs. By contrast, the data used by Bogen (1995) include five chemicals (aniline, carbon disulfide, ethylbenzene, styrene, and toluene) measured on hands (with possibly higher permeability coefficients than skin from other anatomic regions as discussed in Section 6.2.3), four chemicals (2-butoxyethanol, chloroform, trichloroethylene, and tetrachloroethylene) measured in guinea pigs (which are often reported to exhibit larger permeability coefficients than human skin), and the dermal absorption of one chemical (chloroform) estimated indirectly from appearance in breath (which could easily lead to uncertainties of at least an order of magnitude). Finally, Vecchia (1997) has found that experimental measurements of permeability coefficient can deviate from correlations of the form listed in Table 6.2 by at least one order of magnitude. That is, correlation predictions within one order of magnitude of experimental observations are probably not statistically different.

6.4   MATHEMATICAL MODELS OF PENETRATION

Mathematical models of dermal penetration are used for data analyses and for prediction. The choice of model depends on its intended use. For estimating risk from long-term, low-level exposures, frequently only the total amount absorbed during some period is needed (e.g., a lifetime exposure). In this case, a model that predicts the total amount absorbed for a given exposure may be adequate. Probably blood concentration levels remain low and nearly all of the chemical that absorbs into the skin eventually is absorbed systemically.

By contrast, blood concentration levels or concentrations at some target organ are required to estimate risk from acute exposures. In this case, more detailed information on the rate of dermal absorption into the systemic circulation must be available and combined with information on distribution to other tissues and kinetics of metabolism and elimination. That is, a pharmacokinetic model is necessary. Pharmacokinetic models are also needed to determine skin absorption parameters (e.g., permeability coefficients or partition coefficients) from *in vivo* measurements of chemical in blood, exhaled breath, urine, or feces.

6.4.1   MEMBRANE MODELS

Several investigators have solved Fick's law combined with the non-steady-state differential mass balance representing skin penetration as a passive, pseudohomogenous membrane with the properties of the *sc*. If the concentration of the vehicle remains constant, and the systemic concentration remains small, dermal absorption of chemicals can be divided into two periods, as discussed earlier: non-steady-state and steady-state (see Figure 6.1). The complete membrane equation for this case involves an infinite series of terms that are sometimes computationally inconvenient. Cleek and Bunge (1993) have shown that the complete solution can be adequately represented as two simple functions separately representing the non-steady-state and steady-state periods.

During the non-steady-state period, while the chemical is filling the capacity of the sc, the total mass entering the sc (i.e., the absorbed dose) is proportional to $\sqrt{t_{exp}}$:

$$M_{in} = A\,C_w^o \sqrt{\frac{4\left(R_{sc/w} L_{sc}\right) P_{sc}\, t_{exp}}{\pi}} \qquad \text{for } t_{exp} < 2.4\, t_{lag,sc} \qquad (6.15)$$

where A is the exposed area and C_w^o is the concentration of solute in the aqueous solution. The quantity $R_{sc/w}\,L_{sc}$ represents the skin's capacity for the penetrating chemical and can be estimated by multiplying separate values or estimates for $R_{sc/w}$ and L_{sc}. Alternatively, this quantity can be calculated directly by knowing the total mass absorbed into the skin when equilibrated with a solution of concentration C_w^o (i.e., M_{sc}^{eq}), or by knowing the total mass in the skin once steady-state is established with a solution of concentration C_w^o (i.e., M_{sc}^{ss}):

$$R_{sc/w} L_{sc} = \frac{M_{sc}^{eq}}{A\,C_w^o} = \frac{2\,M_{sc}^{ss}}{A\,C_w^o} \qquad (6.16)$$

where A is the external surface area of the exposed skin. Grouped in this way, it is possible to estimate the absorbed dose using readily determined experimental values (C_w^o, A and either M_{sc}^{eq} or M_{sc}^{ss}).

Once the capacity of the sc is filled, steady-state is reached and the absorbed dose increases linearly with time:

$$M_{in} = A\,C_w^o \left[P_{sc} t_{exp} + \frac{\left(R_{sc/w} L_{sc}\right)}{3} \right] \qquad \text{for } t_{exp} > 2.4\, t_{lag,sc} \qquad (6.17)$$

However, the sc capacity for chemical that was filled during the non-steady-state period continues to contribute, as indicated by the right-hand term in Equation (6.17). The time to reach steady-state is approximately 2.4 times $t_{lag,sc}$. Equations (6.15) and (6.17) provide an excellent approximation of the rigorous and much more complex solution to the non-steady-state differential material balance for an absorbing membrane. Equation (6.15) applies even if the ve contributes significantly to the total permeability of the skin. During the non-steady-state period, chemical has not yet penetrated across the sc to find the additional resistance from the ve.

Solute diffusivity through the sc decreases as molecular weight (MW) increases. Consequently, molecules with $MW >$ approximately 500 to 600 are essentially nonpenetrating, except perhaps through skin appendages to a limited extent. Lacking other experimental data, Equation (6.18) can be used to provide a preliminary estimate for $t_{lag,sc}$ (Bunge and Cleek 1995):

$$t_{lag,sc}(\text{hr}) \approx 0.14\left(10^{0.006\,MW}\right) \qquad (6.18)$$

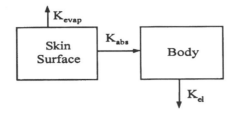

FIGURE 6.9 Schematic of a simple classical pharmacokinetic model which shows two compartments with first-order rates connecting the compartments and describing losses.

in which L_{sc} is approximated as 16 μm (which is consistent with measurements reported by Kalia et al. (1996a)).

Notably, the mass absorbed during both the non-steady-state and steady-state periods depends on the skin capacity to hold chemical (i.e., the skin–water partition coefficient) in addition to the permeability coefficient. Unfortunately, this requirement is often forgotten, and rather limited data are presently available for the skin–water partition coefficient.

6.4.2 PHARMACOKINETIC MODELS

Pharmacokinetic models have wide utility in pharmacology and toxicology to relate external dose to internal dose. Pharmacokinetic models of dermal uptake are generally used to predict absorption, distribution, metabolism, and elimination of chemicals after dermal exposure. They can also be used as tools to estimate the amount of chemical absorbed in the body after *in vivo* penetration studies based on concentrations in breath, blood, urine, or feces. Chemical concentration in tissues and body fluids during and after a dermal exposure is constantly changing but capable of being understood. Pharmacokinetics is the science that describes the behavior (absorption, distribution, metabolism, and elimination) of chemicals in the body. Pharmacokinetics are very important in establishing dose–response relationships. Compartmental descriptions are fundamental to pharmacokinetics.

A simple classical compartmental description is shown in Figure 6.9. Two compartments are shown, each with a loss rate and a one-way transfer rate between them: K_{abs} is the absorption rate from the skin surface to the body; K_{evap} is the evaporation rate from the skin surface; and K_{el} is the elimination rate from the body. The product of one of these first-order rate constants (K_i) and the mass of chemical (M_i) in the compartment i represents the rate of loss or gain of mass in the compartment.

$$\frac{dM_i}{dt} = K_i M_i \tag{6.19}$$

The mass in the body compartment i is directly related to a plasma concentration (C_{plasma}) by a proportionality constant V_i, referred to as the volume of distribution.

$$M_i = V_i C_{plasma} \tag{6.20}$$

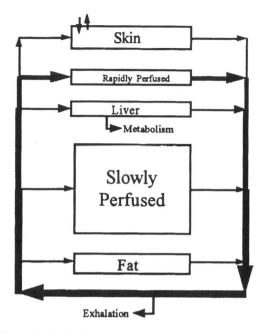

FIGURE 6.10 Schematic of a physiologically based pharmacokinetic showing lumped compartments connected by physiological blood flows.

This proportionality constant has no direct physiologic meaning and does not refer to a real volume. The first-order rates require only simple mathematics to describe the concentration in compartments; for example, the effect of elimination on the concentration in the body compartment i at time t, (i.e., C_i) would be described by:

$$\log C_i = \log C_{i,o} - \frac{K_{el}\, t}{2.303} \tag{6.21}$$

where $C_{i,o}$ is the original concentration in the body compartment i at time zero. The rate of evaporation can be determined independently from the absorption experiment by measuring evaporation from the skin. The rate of elimination can be estimated from an experiment in the same animal where absorption is absent, such as an intravenous exposure. When blood concentration data does not follow simple linear kinetics, additional compartments can be added to make the description more accurate. Numerous studies have used this general approach to estimate the total mass absorbed after a dermal exposure. These linear pharmacokinetic models have the advantage of being fairly straightforward and usable without additional experimental information. A disadvantage is that they are purely descriptive and therefore can only apply when the experimental situation mimics the actual human exposure situation.

Physiologically based pharmacokinetic models have more recently been applied to the task of determining mass absorbed after a dermal exposure. This approach differs from the linear approach in two ways: (1) the recognition that many pharmacokinetic processes are saturable or nonlinear, and (2) the use of more meaningful pharmacokinetic and physiologic parameters. A physiologically based pharmacokinetic model describes the organs of the body as lumped compartments based on the blood flow they receive and their affinity for the chemical of interest. Figure 6.10 shows a schematic of such a model.

The lumped compartments have masses that add up to the total mass of the animal that is perfused with blood. Total blood flow is divided between the compartments based on physiologic measurements of blood flow. Measured partition coefficients, or the ratio of concentrations at equilibrium, between the tissues and the blood are assigned to each lumped compartment. Simultaneous differential equations are used to keep track of the mass in each compartment of the model much like the chemical engineering approach for manufacturing processes. For example, the equation for a lumped compartment, such as fat, with no elimination would be:

$$\frac{dM_i}{dt} = Q_i \left(C_a - \frac{C_i}{R_{i/b}} \right) \tag{6.22}$$

where Q_i is the physiologic flow (L/hr) to tissue compartment i, C_a is arterial blood concentration (mg/L), C_i is the concentration in the tissue compartment i (mg/L), and $R_{i/b}$ is the tissue i to blood partition coefficient (unitless ratio of concentrations). This equation can be compared to Equation (6.19). In this case, several physiologic parameters are used to replace the one first-order rate constant.

Rate constants in a physiologically relevant skin compartment are defined in terms of skin properties such as permeability coefficients, partition coefficients, and skin thickness. A compartment model represents the average concentration in skin as a function of time. A common representation for the skin compartment would be:

$$\frac{dM_m}{dt} = Q_{sk} \left(C_a - \frac{C_m}{R_{m/b}} \right) + P_m A \left(C_w - \frac{C_m}{R_{m/w}} \right) \tag{6.23}$$

in which the permeability coefficient P_m (as given in Equation (6.5)) controls entry of chemical into the skin membrane, Q_{sk} is blood flow to the skin, C_m is concentration in the skin membrane ($= M_m/V_m = M_m/(L_m A)$, where M_m is the mass of chemical in the skin and L_m is the skin thickness), $R_{m/b}$ is the skin-to-blood partition coefficient, A is the surface area of skin exposed, C_w is the concentration of chemical in water on the skin surface, and $R_{m/w}$ is the skin-to-water partition coefficient (as defined in Equation (6.10)). Equation (6.23) is one of several different physiologically relevant compartment representations for skin (McDougal et al. 1986; McCarley and Bunge 1998; Reddy et al. 1998). The various equations are related differently to physiologic properties of the skin compartment (e.g., cutaneous blood flow rate, permeability coefficients, partition coefficients, etc.) causing each equation to match a membrane model description in different respects.

In these physiologically based pharmacokinetic models, mass balance is maintained on the whole system and the differential equations for each physiologic compartment (including the skin compartment) are solved simultaneously. Because these models are based on measurable parameters that are physiologically meaningful, they require more effort to develop and validate. The advantage of this type of model is the potential to be predictive, to extrapolate between species, and to use the model for hypothesis testing.

Physiologically based pharmacokinetic models have recently been used to assess the potential for dermal absorption of contaminants from drinking water in showering and bathing scenarios. Rao and Brown (1993) estimated maximum blood, brain, and skin concentrations after a 15-minute shower or 30-minute bath using trichloroethylene-contaminated water. Rao and Ginsberg (1997) used a physiologically based pharmacokinetic model for methyl t-butyl ether (MTBE) to show that a 15-minute shower in water containing 1 mg/L MTBE would result in brain concentrations twofold below the brain concentrations that would be associated with the acute Minimal Risk Level. This type of model, which has the ability to relate contaminated water concentrations to target tissue dose or body burden has a great deal of potential for understanding showering and bathing hazards.

Hybrid models represent a compromise that utilizes physiologically based parameters for the skin compartment but non-physiologically based classic pharmacokinetic models (Figure 6.9) for describing chemical distribution and rates in body compartments. In these hybrid models, systemic pharmacokinetic parameters must be determined experimentally (usually separately from the dermal absorption experiment) and would not be transferable to a different species. However, independently determined or estimated skin parameters could be used in the physiologically based skin compartment. Hybrid models can be an alternative when insufficient information is available to develop a complete physiologically based pharmacokinetic model.

6.5   RECOMMENDATIONS OF EXPERIMENTAL METHODS FOR WATER EXPOSURES

Dermal absorption is experimentally measured and reported in a wide variety of ways. Common sense has to be used when there are choices about the type of information used in assessing the risk from dermal exposure. Each study has its own merits based on the experimental design, measurements taken, and interpretation of the results. Nevertheless, certain experimental designs, types of measurements, and types of results are preferred.

For the purposes of evaluating showering and bathing scenarios, the study must have the chemical of interest presented in water as a vehicle. Properly designed and analyzed *in vivo* studies would be preferred over *in vitro* studies because the *in vivo* studies (with their blood flow, immune and metabolic responses) are more representative of the human exposure situation. Obviously, human studies would be preferred over laboratory animal studies where possible. Where human studies are not possible, primates would generally be more representative of humans than rodents and swine.

Consideration should be taken of the anatomic differences in skin and their potential impact when choosing species (i.e., hair density, skin thickness, etc.). Most chemicals appear to penetrate through rodent skin more easily than human skin and would be expected to overestimate the risk by a factor of 2 to 3 (e.g., see Vecchia 1997, McDougal et al. 1990).

Appropriate measurements of penetration depend on the experimental design, but generally, with *in vivo* experiments, the more measurements the better. Complete attempts to account for the chemical absorbed, eliminated, lost, and remaining in the skin will give confidence in the results. Measurements of parent chemical and metabolites are preferable to radioactivity studies, when sufficient sensitivity in the biochemical assay can be achieved. If radioactivity is used, care should be taken to distinguish if any portion of the radioactivity is metabolite.

Permeability coefficients are the most useful representation of measurements from aqueous solutions. Fluxes would be next best because they can be converted to permeability coefficients by dividing the flux by the exposure concentration. Flux alone is only descriptive of the particular aqueous exposure concentration used. The laboratory measurements "percent absorbed" and "lag time" alone are not helpful for estimating total absorption in human exposure scenarios. *In vivo* experiments measuring absorption by direct measurements of chemical in skin, blood, and tissues, particularly as a function of time, are generally preferred over indirect determinations of total amount absorbed in a single time period of exposure. However, a mathematical model is required to estimate the total amount absorbed or the permeability coefficient from measurements of chemical in either skin, blood, or other tissues. As a consequence, the quality of the estimate for the total amount absorbed or permeability coefficients from this type of *in vivo* data are tied to the quality of the mathematical model. Mathematical models are not created equal, and different mathematical models can produce very different permeability coefficients from the same set of data. Biologically based models are preferable to descriptive (linear pharmacokinetic) models when laboratory animals are used, because of the potential for extrapolation to other exposure concentrations and durations.

Often it will not be possible to conduct human studies on chemicals of human health concern. Because of the difficulties of extrapolating between animal and human skins, some of the advantages of *in vivo* experiments are lost. Consequently, *in vitro* diffusion cell measurements are valuable. Another advantage of *in vitro* diffusion cells experiments is that the procedures for determining the permeability coefficients and other parameters are well established.

In the preferred *in vitro* configuration, the donor chamber should consist of a well-stirred aqueous solution in direct contact with human skin. Some researchers argue that freshly excised skin is preferred over cadaver skin, although this has not been conclusively demonstrated, except when chemicals are metabolized in the viable epidermis. For most chemicals of concern in drinking water exposures, the *sc* is the primary resistance, and the tissue of most interest. Standard procedures for preparing *sc*-only samples are well established. When possible, skin integrity should be established by confirming that the permeability coefficient of tritiated water does not exceed 1.5×10^{-3}.

The receptor solution should be stirred and configured to keep the concentration of the penetrating chemical low (i.e., to maintain sink conditions), either by periodic or continuous replacement or use of a large receptor volume relative to the receptor fluid capacity for the penetrating chemical. The experiment should be conducted long enough so that the rate of mass appearing in the receptor fluid is constant. When calculating the permeability coefficient, only receptor fluid data collected for times longer than about 2.4 times the lag time should be used. Temperature does affect penetration rates to some degree and a physiologically relevant temperature such as 32°C is preferred. The receptor fluid solution should be principally water, and additives designed to increase chemical solubility must not damage or alter the skin. The concentration of the dissolved, unionized species of the chemical must be known. Consequently, the pH should be measured for weak acids or bases.

Correlation approaches to determine permeability provide a larger perspective on the problem when properly applied to a large group of individual studies. They are very useful, especially when limited information is available on the chemical of interest.

6.6 RECOMMENDATIONS FOR USE OF PENETRATION DATA FOR RISK ASSESSMENTS

The physical form of chemical exposure to the skin in the showering and bathing scenarios could be either vapor, aerosol, or aqueous solution. Absorption from aqueous solutions is by far the most important route for dermal exposure to drinking water. Vapor and aerosol exposures to the skin can be ignored based on the rationale below. Only low vapor concentrations would be achieved from the small amount of contamination in water. Vapor exposures need only be considered where the skin is dry. When the skin is wet due to showering or bathing, the chemical exposure would tend to be via an aqueous vehicle. Pulmonary absorption is a much more efficient route for absorption of vapors than the skin; where a high vapor concentration exists in showering and bathing scenarios, vapor absorption through the skin would be insignificant. Aerosol exposures to the skin would be from aerosolized water droplets containing the chemical. These aerosol exposures would ultimately be aqueous exposures when the aerosol contacts the skin. Contaminated water that was aerosolized and subsequently deposited on the skin will generally contact a much smaller skin area than liquid water. In the case of volatile chemicals, these aerosols would probably have a lower contaminant concentration than the water that had not aerosolized. Therefore, the aerosol form of the contaminated water would not need a separate assessment.

The primary absorption concern in the showering and bathing scenario would be the direct exposure to contaminated water. Dermal absorption data from exposures to chemicals in aqueous solutions should be used with appropriate estimates of body surface area and exposure area to estimate the total chemical absorbed.

Skin condition, superhydration of the skin, high temperatures, and the use of soaps and detergents are other factors potentially present in bathing and showering exposures that were not present in laboratory studies used to estimate dermal absorption. Damaged skin might impact the dermal uptake of chemicals from contaminated water if the ratio of damaged surface area to total surface area exposed becomes significant. Localized dermatitis or cracked skin on hands or feet would probably add only a minimal amount to the total exposure. For the purpose of estimating general hazards from showering and bathing, it is probably safe to assume skin is undamaged. Superhydration of the skin would not be expected to be a consideration during the 10- to 15-minute showering and bathing scenario, but might be important during extended bathing. The direct effect of temperature on the penetration of the chemical, by changing the fluidity characteristics of the skin, has been shown only experimentally after several hours at elevated temperature. Increased temperature might be expected to increase the blood flow significantly during the short showering and bathing scenario. For those chemicals that penetrate to the extent that blood flow to the skin is the limiting factor, rather than diffusion through the skin, penetration could be enhanced when blood flow is increased. The impact of blood flow on total dose absorbed and the types of chemicals for which blood flow may be of concern need more study. The effect of household soap on dermal penetration during a normal showering and bathing scenario needs to be studied. There is not enough information to quantitatively assess the effect of these factors on skin penetration. For volatile chemicals, it is anticipated that none of these changes would have a significant impact on the body burden. Except for premature infants within 4 weeks of birth, estimates of parameters representing skin barrier function should apply to most individuals (including special populations such as pregnant women or children).

Given appropriate permeability coefficients and equilibrium coefficients, it is fairly easy to develop estimates of body burden from a dermal exposure. For a particular exposure scenario where exposed surface area, aqueous concentration, and duration of exposure can be estimated, mass absorbed can be calculated according to Equations (6.15) or (6.17) combined with Equation (6.18) and the assumption that the skin (actually sc) thickness is about 16 μm. If necessary, Equations (6.11) and (6.13) can be used to estimate permeability and partition coefficient values required in Equations (6.15) and (6.17). Chemical concentration should be the dissolved concentration of only the unionized species if the chemical ionizes. Chemicals that are almost completely ionized at exposure conditions (e.g., less than 1% of the dissolved chemical is unionized) can be absorbed, albeit much more slowly than the unionized species (as defined by Equations (6.9) and (6.14)). Adverse systemic health effects from dermal absorption of an ionized chemical are unlikely unless the level of contamination is quite high and the chemical is highly toxic. Finally, the use and interpretation of these methods in exposure assessment should recognize that all estimates of dermal uptake may involve significant uncertainties, as much as an order of magnitude and perhaps more for certain classes of chemicals such as organohalides.

REFERENCES

Abraham, M.H., Chadha, H.S., and Mitchell, R.C. (1995). The factors that influence skin penetration of solutes. J. Pharm. Pharmacol. 47:8–16.

Anderson, B.D., Higuchi, W.I., and Raykar, P.V. (1988). Heterogeneity effects on permeability–partition coefficient relationships in human sc. Pharm. Res. 5:566–573.

Auclair, F., Besnard, M., Dupont, C., and Wepierre, J. (1991). Importance of blood flow to the local distribution of drugs after percutaneous absorption in the bipediculated dorsal flap of the hairless rat. Skin Pharmacol. 4:1–8.

Barry, B.W., Harrison, S.M., and Dugard, P.H. (1984). Correlation of thermodynamic activity and vapor diffusion through human skin for the model compound, benzyl alcohol. J. Pharm. Pharmacol. 37:84–90.

Behl, C.R., Flynn, G.L., Kurihara, T., Harper, N., Smith, W., Higuchi, W.I., Ho, N.F., and Pierson, C.L. (1980). Hydration and percutaneous absorption: I. Influence of hydration on alkanol permeation through hairless mouse skin. J. Invest. Dermatol. 75:346–352.

Berner, B. and Cooper, E.R. (1987). Models of skin permeability. In: Transdermal Delivery of Drugs, Vol. II (A.F. Kydonieus and B. Berner, Eds.), pp. 41–56. CRC Press, Boca Raton, FL.

Blank, I.H. (1953). Further observations on factors which influence the water content of the stratum corneum. J. Invest. Dermatol. 21:259–269.

Blank, I.H. and McAuliffe, D.J. (1985). Penetration of benzene through human skin. J. Invest. Dermatol. 85(6):522–536.

Blank, I.H., Moloney, J.I., Emsile, A.G., Simon, I., and Apt, C. (1984). The diffusion of water across the stratum corneum as a function of its water content. J. Invest. Dermatol. 82(2):188–194.

Blank, I.H. and Scheuplein, R.J. (1964). The epidermal barrier. In: Progress in Biological Sciences in Relation to Dermatology, pp. 245–261, Cambridge University Press, New York.

Blank, I.H., Scheuplein, R.J., and MacFarlane, D.J. (1967). Mechanism of percutaneous absorption. III. The effect of temperature on the transport of non-electrolytes across the skin. J. Invest. Dermatol. 49(6):582–589.

Bogen, K.T. (1995). Models based on steady-state in vitro dermal permeability data underestimate short-term in vivo exposures to organic chemicals in water, J. Expos. Anal. Environ. Epidem. 4: 457–475.

Bronaugh, R.L. and Barton, C.N. (1992). Prediction of human percutaneous absorption with physicochemical data. In: Health Risk Assessment through Dermal and Inhalation Exposure and Absorption of Toxicants. (R.G. Wang and H.I. Maibach, Eds.), CRC Press, Boca Raton, FL.

Bronaugh, R.L., Stewart, R.F. and Simon, M. (1986). Methods for in vitro percutaneous absorption studies. VII. Use of excised human skin. J. Pharm. Sci. 75:1094–1097.

Brown, S.L. and Rossi, J.E. (1989). A simple method for estimating dermal absorption of chemicals in water. Chemosphere 19:1989–2001.

Bunge, A.L. and Cleek, R.L. (1995). A new method for estimating dermal absorption from chemical exposure. 2. Effect of molecular weight and octanol-water partitioning. Pharm. Res. 12:88–95.

Bunge, A.L., Cleek, R.L., and Vecchia, B.E. (1995). A new method for estimating dermal absorption from chemical exposure. 3. Compared with steady-state methods for prediction and data analysis. Pharm. Res. 12:972–982.

Bunge, A.L., Flynn, G.L., and Guy, R.H. (1994). Predictive model for dermal exposure assessment, In: Drinking Water Contamination and Health: Integration of Exposure Assessment, Toxicology, and Risk Assessment, (R.Wang, Ed.), Marcel Dekker, New York, pp. 347–373.

Bunge, A.L. and McCarley, K.D (1996). An improved physiologically relevant pharmacokinetic model for skin. In: Prediction of Percutaneous Penetration. Vol. 4b. (K.R. Brain, V.J. James, and K.A. Walters, Eds.), pp. 190–194, STS Publishing, Cardiff, U.K.

Chang, S.K. and Riviere J.E. (1991). Pecutaneous absorption of parathion *in vitro* in porcine skin: Effects of dose, temperature, humidity, and perfusate composition on absorptive flux. Fund. Appl. Toxicol. 17:494–504.

Cleek, R.L. and Bunge, A.L. (1993). A new method for estimating dermal absorption from chemical exposure. 1. General approach. Pharm. Res. 10:497–506.

Crank, J. (1975). The Mathematics of Diffusion, Oxford University Press, London.

Cussler, E.L. (1984). Diffusion: Mass Transfer in Fluid System, Cambridge University Press, New York.

Dugard, P.H. and Scott, R.C. (1984). Absorption through skin. In: Chemotherapy of Psoriasis (H.P. Baden, Ed.), pp. 125–144. Pergamon Press, Oxford.

Dutkiewicz, T. and Piotrowski, J. (1961). Experimental investigations on the quantitative estimation of aniline absorption in man. Pure Appl. Chem. 3:319–323.

Dutkiewicz, T. and Tyras, H. (1968). The quantitative estimation of toluene skin absorption in man. Int. Archiv. Fur Gwerbepathologie und Gwerbehygiene 24:253–257.

El Tayar, N., Tsai, R.-S., Testa, B., Carrupt, P.-A., Hansch, C., and Leo, A. (1991). Percutaneous penetration of drugs: a quantitative structure-permeability relationship study. J. Pharm. Sci. 80:744–749.

Feldman, R.J. and Maibach, H.I. (1967). Regional variation in percutaneous penetration of ^{14}C cortisol in man. J. Invest. Derm. 48:181–183.

Fiserova-Bergerova, V., Pierce, J.T., and Droz, P.O. (1990). Dermal absorption potential of industrial chemicals: criteria for skin notation. Am. J. Ind. Med. 17:617–635.

Flynn, G.L. (1990). Physicochemical Determinants of Skin Absorption. In: Principles of Route-to-Route Extrapolation for Risk Assessment. (T.R. Gerrity and C.J. Henry, Eds.), pp. 93–127. Elsevier, New York.

Gordon, S., Callahan, P., Brinkman, M., Kenny, K., and Wallace, L. (1997). Effect of water temperature on dermal exposure to chloroform. 7th Annual Meeting of the International Society of Exposure Analysis. Research Triangle Park, NC. Nov. 2–5.

Gromiec, J.P. and Piotrowski, J.K. (1984). Urinary mandelic acid as an exposure test for ethylbenzene. Int. Arch. Occup. Environ. Health 55:61–72.

Hefner, R.E.J., Watanabe, P.G., and Gehring, P.J. (1975). Percutaneous absorption of vinyl chloride. Toxicol. Appl. Pharmacol. 34:529–532.

Idson, B. (1971) Biophysical factors in skin penetration. J. Soc. Cosmet. Chem. 22:615–634.

Jacobs, R.R. and Phanprasit, W. (1993). An *in vitro* comparison of the permeation of chemicals in vapor and liquid phase through pig skin. Adv. Ind. Environ. Hyg. 54(10):569–575.

Jetzer, W.E., Hou, S.Y.E., Huq, A.S., Duraiswamy, N.E.H., and Flynn, G.L. (1988). Temperature dependency of skin permeation of waterborne organic compounds. Pharm. Acta. Helv. 63(7):197–201.

Jo, W.K., Weisel, C.P., and Lioy, P.J. (1990). Routes of chloroform exposure and body burden from showering with chlorinated tap water. Risk Analysis. 10: 575–580.

Kalia, Y.N., Pirot, F., and Guy, R.H. (1996a). Homogeneous transport in a heterogeneous membrane: Water diffusion across human stratum corneum *in vivo*. Biophysical J. 71:2692–2700.

Kalia, Y.N., Nonato, L.B., Lund, C.H., and Guy, R.G. (1996b). Development of skin barrier function in low birthweight infants. Pharm. Res. 13(9): S382.

Kasting, G.B. and Robinson, P.J. (1993) Can we assign an upper limit to skin permeability? Pharm. Res. 10(6):930–931.

Kasting, G.B., Smith, R.L., and Anderson, B.D. (1992). Prodrugs for dermal delivery: solubility, molecular size, and functional group effects. In: Prodrugs: Topical and Ocular Drug Delivery (K.B. Sloan, Ed.), pp. 117–161. Marcel Dekker, New York.

Lambert, W.J., Higuchi,W.I., Knutson, K., and Krill, S.L. (1989). Effects of long-term hydration leading to the development of polar channels in hairless mouse stratum corneum. J. Pharm. Sci. 78(11):925–928.

Leung, H.-W. and Paustenbach, D.J. (1994). Techniques for estimating percutaneous absorption of chemicals due to occupational and environmental exposure. Appl. Occup. Environ. Hyg. 9:187–197.

Lindberg, M., Safstrom, S., Roomans, G.M., and Forslind, B. (1989). Sodium lauryl sulfate enhances nickel penetration through guinea-pig skin. Studies with energy dispersive X-ray microanalysis. Scanning Microsc. 3(1):221–224.

Liron, Z., Clewell, H.J., and McDougal, J.N. (1994). Kinetics of water vapor sorption in porcine stratum corneum. J. Pharm. Sci. 83(5):692–698.

Maibach, H.I., Feldman, R.J., Milby, T.H., and Serat, W.F. (1971). Regional variation in percutaneous penetration in man. Arch. Environ. Health 23:208–211.

McCarley, K.D. and Bunge, A.L. (1998). Physiologically relevant one-compartment pharmacokinetic models for skin. 1. Development of models. J. Pharm. Sci. 87:470–481.

McCollister, D.D., Beamer, W.H., Atchison, G.J., and Spencer, H.C. (1951). The absorption, distribution and elimination of radioactive carbon tetrachloride by monkeys upon exposure to low vapor concentrations. J. Pharmacol. Exp. Ther. 102:112–124.

McDougal, J.N., Jepson, G.W., Clewell, H.J., III, and Andersen, M.E. (1985). Dermal absorption of dihalomethane vapors. Toxicol. Appl. Pharmacol. 79:150–158.

McDougal, J.N., Jepson, G.W., Clewell, H.J., III, Gargas, M.L., and Andersen, M.E. (1990). Dermal absorption of organic chemical vapors in rats and humans. Fundam. Appl. Toxicol. 14:299–308.

McDougal, J.N., Jepson, G.W., Clewell, H.J., III, MacNaughton, M.G., and Andersen, M.E. (1986). A physiological pharmacokinetic model for dermal absorption of vapors in the rat. Toxicol. Appl. Pharmacol. 85:286–294.

McKone, T.E. and Howd, R.A. (1992). Estimating dermal uptake of nonionic organic chemicals from water and soil: I. Unified fugacity-based models for risk assessment. Risk Anal. 12:543–557.

Michaels, A.S., Chandrasekaran, S.K., and Shaw, J.E. (1975). Drug permeation through human skin: theory and in vitro experimental measurement. AIChE J. 21:985–996.

Morimoto, Y., Hatanaka, T., Sugibayashi, K., and Omiya, H. (1992). Prediction of skin permeability of drugs: comparison of human and hairless rat skin. J. Pharm. Pharmacol. 44:634–639.

Parry, G.E., Bunge, A.L., Silcox, G.D., Pershing, L.K., and Pershing, D.W. (1990). Percutaneous absorption of benzoic acid across human skin. I. In vitro experiments and mathematical modeling. Pharm. Res. 7:230–236.

Piotrowski, J.K. (1971). Evaluation of exposure to phenol: absorption of phenol vapor in the lungs and through the skin excretion of phenol in urine. Br. J. Ind. Med. 28:172–178.

Pirot, F., Kalia, Y.N., Stinchcomb, A.L., Keating, G., Bunge, A., and Guy, R.H. (1997). Characterization of the permeability barrier of human skin in vivo. Proc. Natl. Acad. Sci. USA, 94:1562–1567.

Potts, R.O. and Guy, R.H. (1993). Predicting skin permeability. Pharm. Res. 9:663–669.

Rao, H.V. and Brown, D.R. (1993). A physiologically based pharmacokinetic assessment of tetrachloroethylene in groundwater for a bathing and showering determination. Risk Anal. 13(1):37–49.

Rao, H.V. and Ginsberg, G.L. (1997). A physiologically-based pharmacokinetic model assessment of methyl t-butyl ether in groundwater for a bathing and showering determination. Risk Anal. 17(5):583–598.

Reddy, M.B., McCarley, K.D., and Bunge, A.L. (1998). Physiologically relevant one-compartment pharmacokinetic models for skin. 2. Comparison of models combined with a systemic pharmacokinetic model. J. Pharm. Sci. 87:482–490.

Riihimaki, V. and Pfaffli, P. (1978). Percutaneous absorption of solvent vapors in man. Scand. J. Environ. Health 4:73–85.

Riviere, J.E. and Williams, P.L. (1992). Pharmacokinetic implictions of changing blood flow in the skin. Letter to the Editor, J. Pharm. Sci. 81(6):601–602.

Roberts, M.S., Anderson, R.A., Moore, D.E., and Swarbrick, J. (1977). The distribution of nonelectrolytes between human stratum corneum and water. Aust. J. Pharm. Sci. 6:77–82.

Roberts, M.S., Anderson, R.A., and Swarbrick, J. (1977). Permeability of human epidermis to phenolic compounds. J. Pharm. Pharmacol. 29:677–683.

Roy, S.D. and Flynn, G.L. (1990). Transdermal delivery of narcotic analgesis: pH, anatomical and subject influences on cutaneous permeability of fentanyl and sufentanil, Pharm. Res. 7:842–847.

Schaefer, H., Zesch, A., and Stuttgen, G. (1982). Skin Permeability, Springer-Verlag, Berlin and New York.

Scheuplein, R.J. (1967a) Mechanism of percutaneous absorption. II. Transient diffusion and the relative importance of various routes of skin penetration. J. Invest. Dermatol. 48:79–88.

Scheuplein, R.J. (1967b). Molecular structure and diffusional process across intact skin. Final report to the U.S. Army Chemical RandD Laboratories, Edgewood Arsenal, MD, Contract DA 18-108-AMC-726(A).

Scheuplein, R.J. and Blank, I.H. (1967). Molecular structure and diffusional processes across intact skin. Report to the U.S. Army Chemical R&D Laboratories, Edgewood Arsenal, MD.

Scheuplein, R.J. and Blank, I.H. (1971). Permeability of the skin. Physiol. Rev. 51:702–747.

Scheuplein, R.J. and Blank, I.H. (1973). Mechanism of percutaneous absorption. IV. Penetration of non-electrolytes (alcohols) from aqueous solutions and from pure liquids. J. Invest. Dermatol. 60:286–296.

Scheuplein, R.J., Blank, I.H., Brauner, G.J., and MacFarlane, D.J. (1969). Percutaneous absorption of steroids. J. Invest. Dermatol. 52:63–70.

Scheuplein, R.J. and Bronaugh, R.L. (1983). Percutaneous absorption. In: Biochemistry and Physiology of the Skin, Vol. 2 (L.A. Goldsmith, Ed.), pp. 1255–1295. Oxford University Press, Oxford.

Scheuplein, R.J. and Ross, E.W. (1974). Mechanism of percutaneous absorption. V. Percutaneous absorption of solvent deposited solids. J. Invest. Dermatol. 62(4):353–360.

Siddiqui, O., Roberts, M.S., and Polack, A.E. (1989). Percutaneous absorption of steroids: relative contributions of epidermal penetration and dermal clearance. J. Pharmacokin. Biopharm. 17:405–424.

Sprott, W.E. (1965). Surfactants and percutaneous absorption. Transcripts St. John's Hosp. Dermatol. Soc. 51:56–71.

Surber, C., Wilhelm, K.-P., Hori, M., Maibach, H.I., and Guy, R.H. (1990a). Optimization of topical therapy: partitioning of drugs in stratum corneum. Pharm. Res. 7:1320–1324.

Surber, C., Wilhelm, K.-P., Maibach, H.I., Hall, L.L., and Guy, R.H. (1990b). Partitioning of chemicals into human stratum corneum: Implications for risk assessment following dermal exposure. Fund. Appl. Toxicol. 15:510–516.

Tsuruta, H. (1989). Skin absorption of organic solvent vapors in nude mice *in vivo*. Ind. Health 27:37–47.

US EPA (1992). Exposure Assessment Group, Office of Research & Development, Interim Report, EPA/600/8-91/011B, Dermal Exposure Assessment: Principles and Application.

Van der Valk, P.G., Nater, J.P., and Bleumink, E. (1984). Skin irritancy of surfactants as assessed by water vapor loss measurements. J. Invest. Dermatol. 82(3):291–293.

Vecchia, B.E. (1997). Estimating Dermal Absorption: Data Analysis, Parameter Estimation and Sensitivity to Parameter Uncertainties, M.S. Thesis, Colorado School of Mines, Golden, Colorado.

Wade, O.L., Bishop, J.M., and Donald, K.W. (1962). Cardiac Output and Regional Blood Flow. F.A. Davis, Philadelphia, PA.

Walker, J.D., Whittaker, C., and McDougal, J.N. (1996). Role of the TSCA Interagency Testing Committee in meeting the U.S. Government data needs: designating chemicals for percutaneous absorption testing. In: Dermatotoxicology, 4th edition, volume 2, pp. 371–381, Hemisphere Publishing New York, Washington, Philadelphia, and London.

Weisel, C.P. and Shepard, T.A. (1994). Chloroform exposure and the body burden associated with swimming in chlorinated pools. In: Water Contamination and Health: Integration of Exposure Assessment, Toxicology, and Risk Assessment. (R. Wang, Ed.), pp. 135–148. Marcel Dekker, New York.

Wieczorek, H. (1985). Evaluation of low exposure to styrene: II. Dermal absorption of styrene vapours in humans under experimental conditions. Int. Arch. Occup. Environ. Health 57:71–75.

Wilhelm, K., Surber, C., and Maibach, H.I. (1991). Effect of sodium lauryl sulfate-induced skin irritation on in vivo percutaneous penetration of four drugs. J. Invest. Dermatol. 97(5):927–932.

7 Case Study

Charles R. Wilkes

CONTENTS

1-57881-001-9/99/$0.00+$.50
© 1999 by International Life Sciences Institute

183

7.1 INTRODUCTION

The purpose of this case study is to demonstrate an application of the methods described and recommended in this book. This case study uses appropriate available information and techniques to represent population exposure to three potential water contaminants and to assemble an estimate of the range and shape of the distribution of exposures to the population. The three water contaminants investigated in this case study were chloroform, methyl parathion, and chromium, chosen to represent a wide range of volatilities, solubilities, and other chemical properties. The water supply was assumed to be contaminated with each contaminant at a concentration of 1 mg/L.

The exposure is estimated on a case-by-case basis using an indoor air quality and human exposure model (MAVRIQ; see Wilkes, 1994 and Wilkes et al., 1996) specifically adapted for this purpose. MAVRIQ is a mass-balance model that uses finite difference techniques to solve the system of differential equations. For all cases, a typical house is used, with the housing characteristics (i.e., ventilation rates, configuration, etc.) held constant for all simulations. Occupant locations within the home are sampled from a human activity pattern database. Depending on the locations, appropriate water uses are simulated, reflecting survey results of human water use characteristics. By accounting for water uses, contaminant releases from these water uses, and subsequent contaminant transport throughout the residence, the model predicts indoor air concentrations at various locations in the home. Combining these concentrations with occupant locations, the model calculates the occupant's exposure to the contaminant. And finally, the model translates this data using a pharmacokinetic calculation and estimates the quantity of contaminant actually absorbed by the body through both inhalation and dermal routes.

Numerous simulations are executed, resulting in an estimated distribution of absorbed contaminant dose for the given population group. The population group evaluated in this case study is sampled from the Survey of Activity Patterns of California Residences (CARB Activity Pattern Database).

Each part of the process is described in detail in the following sections.

7.2 MODELED RESIDENCE

The modeled residence is an idealized single-story, one-bedroom house, as shown in Figure 7.1. The airflows indicated in Figure 7.1 are assigned to represent airflows typically found in U.S. housing, with a whole-house air exchange rate of 0.5 air exchanges/hour (ACH). Two air flow regimes are defined, representing likely physical changes in the residence. The first regime represents normal conditions with all interior doors open. The second (shown in brackets) represents the condition when someone is occupying the bathroom or shower and corresponds to the physical change of closing the bathroom door. The airflows between the bathroom and hallway are approximately two thirds smaller when the door is closed. The locations of the water-use devices available for use by the occupants are also shown in Figure 7.1.

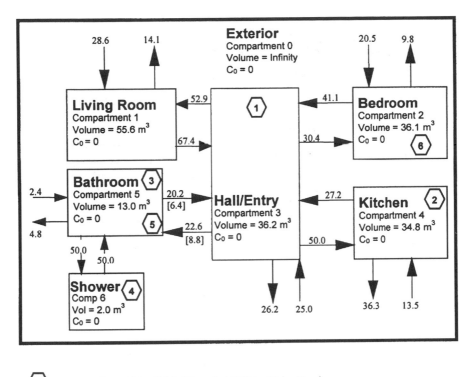

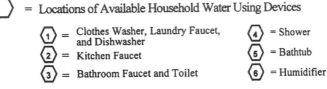

= Locations of Available Household Water Using Devices

⟨1⟩ = Clothes Washer, Laundry Faucet,   ⟨4⟩ = Shower
      and Dishwasher
⟨2⟩ = Kitchen Faucet   ⟨5⟩ = Bathtub
⟨3⟩ = Bathroom Faucet and Toilet   ⟨6⟩ = Humidifier

FIGURE 7.1 Compartmental representation of the modified IBR test house (After Axley, 1988). Airflow rates assigned such that the whole house air exchange rate = 0.5 ACH. Flows are in m³/hour. The flows shown in brackets [] are in effect when the bathroom door is closed.

7.3 ACTIVITY PATTERNS

The activity patterns are sampled from the Survey of Activity Patterns of California Residences (CARB Activity Pattern Database). The database contains 24-hour records of the activities and locations for each individual recorded in the database. Each activity recorded in the database has a start time, an end time, and two codes, one representing the location and one representing the activity during the time period. The list of possible location and activity codes and their meanings are given in Tables 7.1 and 7.2, respectively.

The household chosen for this case study is comprised of two adults: one female and one male, each between the ages of 18 and 34. The sampled activity patterns

TABLE 7.1
Locations Recorded in the Survey of California Activity Patterns

Location Code	Location Name	Location Description
001	Home–Bedroom	Bedrooms in a residential home
002	Home–Living	Living areas in a residential home
003	Home–Kitchen	Kitchens in a residential home
004	Home–Dining	Dining areas in a residential home
005	Home–Bathroom	Bathrooms in a residential home
006	Home–Office	Office rooms in a residential home
007	Home–Garage	Garages in a residential home
008	Home–Basement	Basement in a residential home
009	Home–Laundry	Laundry areas in a residential home
010	Home–Pool	Pool areas of a residential home
011	Home–Transit	Moving around a residential home
012	Home–Other	Other areas in a residential home
013	Home–Out	Outside a residential home
014	Office	Office areas at work place
015	Factory	Factory area
016	Grocery Store	Grocery stores
017	Mall	Shopping Malls
018	School	Schools and Universities
019	Public Building	Public buildings
020	Hospital	Hospitals and medical service facilities
021	Restaurant	Restaurants
022	Bar	Bars and night clubs
023	Church	Churches, Chapels, Cathedrals, etc.
024	Gym/Health Club	Gyms and health clubs
025	Other Home	Other residential homes
026	Auto Shop	Auto shops and car repair garages
027	Hotel	Hotels, including related facilities
028	Cleaners	Cleaner shops, laundry facilities, etc.
029	Beauty Parlor	Parlors in residential homes
030	Work–Transit	Transition to and from work
031	Other Indoors	Other indoor areas
032	Out–Park	Open-air areas in parks
033	Other Outdoors	Other outdoor areas
034	Transit	Transit

for both adults are translated from the original domain (the sampled individual's home and work environment) into the model domain (the modeled house and exterior), and subsequently analyzed for inhalation and dermal exposure and uptake. For each simulation, an activity pattern for each adult is sampled from the database. The sampled locations are then mapped to the appropriate location in the modeled residence, as specified in Table 7.3.

TABLE 7.2
Activities Recorded in the Survey of California Activity Patterns

Activity	Activity Description	Activity	Activity Description
1	Work — main job	49	Personal care — travel
2	Work — unemployed	50	Education — student class
3	Work — travel during	51	Education — other class
5	Work — second job	54	Education — homework
6	Work — eat	55	Education — library
7	Work — before/after	56	Education — other
8	Work — break	59	Education — travel to/from
9	Work — travel to/from	60	Org. — professional/union
10	Hh work — prepare food	61	Org. — special interest
11	Hh work — clean-up after meal	62	Org. — political/civil
12	Hh work — clean house	63	Org. — volunteer/help
13	Hh work — clean outdoors	64	Org. — religious group
14	Hh work — clothes care	65	Org. — religious practice
15	Hh work — car repair/maintain	66	Org. — fraternal
16	Hh work — other repairs	67	Org. — child/youth/family
17	Hh work — plant care	68	Org. — other
18	Hh work — animal care	69	Org. — travel to/from
19	Hh work — other	70	Social — sport event
20	Baby care	71	Social — entertainment event
21	Child care	72	Social — movie
22	Child — help/teach	73	Social — theatre
23	Child — read/talk	74	Social — museum
24	Child — play with (indoor)	75	Social — visit
25	Child — play with (outdoor)	76	Social — party
26	Child — medical care	77	Social — bar/lounge
27	Child — other care	78	Social — other
28	At dry cleaners	79	Social — travel to/from
29	Child care — travel	80	Recreation — active sport
30	Grocery shopping	81	Recreation — outdoor
31	Durable shopping	82	Recreation — walk/hike
32	Personal services	83	Recreation — hobby
33	Medical appointments	84	Recreation — domestic craft
34	Govt/financial services	85	Recreation — art
35	Car repair services	86	Recreation — music/drama/dance
36	Other repair services	87	Recreation — games
37	Other services	88	Recreation — computer use
38	Errands	89	Recreation — travel
39	Goods and services — travel	90	Communication — radio
40	Personal — wash, etc.	91	Communication — television
41	Personal — medical care	92	Communication — records/tapes
42	Personal — help and care	93	Communication — read books
43	Personal — meals at home	94	Communication — read magazines
44	Personal — meals out	95	Communication — read newspaper
45	Personal — night sleep	96	Communication — conversations
46	Personal — day sleep/nap	97	Communication — write
47	Personal — dress, etc.	98	Communication — think, relax
48	Personal — n/a activities	99	Communication — travel

TABLE 7.3
Criteria Used for Mapping Activities from the Original Domain (as recorded in the activity pattern database) to the Model Domain (the modeled house and exterior)

Location Code		Activity Code		Model[a]
Code	Description	Code	Description	Compartment
1	Home — Bedroom	40	Personal Care — Wash[b]	5
1	Home — Bedroom	0	All others	2
2	Home — Living Room	40	Personal Care — Wash[b]	5
2	Home — Living Room	0	All others	1
3	Home — Kitchen	40	Personal Care — Wash[b]	5
3	Home — Kitchen	0	All others	4
4	Home — Dining Room	40	Personal Care — Wash[b]	5
4	Home — Dining Room	0	All others	4
5	Home — Bathroom	0	All others	5
6	Home — Office	0	All others	1
9	Home — Laundry	0	All others	3
11	Home — Other	0	All others	3
12	Home — Transit	0	All others	3
0	All others	0	All others	0

[a] See Figure 7.1
[b] All activities of personal care-wash were mapped to the bathroom (compartment 5).

7.4   WATER USES

The activity patterns recorded in the database have little information about the water uses that occurred during the recorded activities. For this reason, the water-use activities are simulated based on characteristic water use information obtained in separate surveys. The variability in activity patterns has been shown to have a significant impact on exposure. The method for simulation of water uses, presented below, attempts to capture the variability that is inherent between people by using actual activity patterns sampled from an activity pattern database and simulating water uses appropriate to the activity pattern.

The water-use characteristics are based on several surveys that are summarized in the *Exposure Factors Handbook* (U.S. EPA, 1997, Chapter 16). The primary source of information for event durations and frequencies is taken from the study by the U.S. Department of Housing and Urban Development (U.S. DHUD, 1984), which monitored over 200 households for a 20-month period. Several of the smaller studies also discussed in the *Exposure Factors Handbook* were used. The values reported in the *Exposure Factors Handbook* are the best available published information; however, much of the data are more than 10 years old and may not reflect current water-use patterns.

The characteristics used to simulate the occupant water uses for this case study are given in Table 7.4. The values given in Table 7.4 for event duration for the

TABLE 7.4
Water Use Characteristics Used as Model Inputs for each Individual

Water Source	Flowrate (liters/min)	Volume of Water Used (liters/event)	Event Duration Mean (min/event)	Event Duration Standard Deviation (min/event)	Event Frequency (Events/day)
Shower	10	—	6.9	5.0	0.74
Bath	—	190	12.0	5.0	0.14
Toilet	—	17	Until next event	0	3.3
Dishwasher	—	41	73.0	6.8	0.4
Clothes Washer	—	188	45.0	6.0	0.29
Faucet (Kitchen)	5.28	—	0.5	0.3	4.0
Faucet (Bathroom)	5.28	—	0.5	0.3	4.0
Faucet (Laundry)	5.28	—	0.5	0.3	4.0

dishwasher and clothes washer are consistent with those reported in the *Exposure Factors Handbook* in Tables 16.20 and 16.22, respectively. In response to customer demand and Department of Energy (DOE) rules, the water consumption in new machines has decreased significantly from the values reported in the *Exposure Factors Handbook* (Hakkinen, 1997). New dishwashers currently average between 8.5 and 9 gallons per use as compared to the range of 9.5 to 12 gallons per use reported in the *Exposure Factors Handbook*. The weighted average water use of new clothes washers is 35.6 gallons per cycle compared to the range of 41 to 55 gallons reported in the *Exposure Factors Handbook* (U.S. EPA, 1997, Chapter 16). Approximately 60% of the machines in use are "extra large/super capacity" washers averaging about 39.2 gallons per cycle and 40% of the machines in use are "large capacity" washers using approximately 30.2 gallons per cycle (Hakkinen, 1997). Ideally, any estimate of average water use per cycle would be calculated from the frequency of use for each machine size, but this is not possible as frequency information is not currently available.

The water uses are modeled by first identifying eligible periods in the sampled activity pattern, then occurrences are simulated based on a Poisson process, and finally durations are simulated based on a lognormal distribution. This process is described in detail in the discussion below and in the schematic representation of this process given in Figure 7.2.

7.4.1 SIMULATING WATER USE OCCURRENCES

Water-use occurrences are simulated as a Poisson process using the information provided in the water-use file and the sample activity pattern shown schematically in Figure 7.2. The location and activity codes are used to identify eligible time periods during the day when each water-use activity is eligible to occur, as described earlier and specified in Table 7.3. For example, if the occupant is in the kitchen, the kitchen faucet water-use activity is eligible to occur.

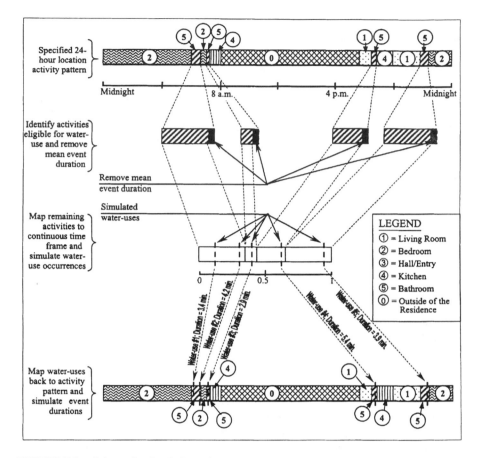

FIGURE 7.2 Schematic simulation of water-use activities based on activity pattern. As an illustrative example, this figure presents the procedure implemented to simulate bathroom water-uses. Note: Activity pattern and simulated water use occurrences and durations are for illustrative purposes only and do not reflect actual events.

The locations are mapped in a hierarchical order. When a particular activity and location code combination matches more than one entry in Table 7.3, the entry is mapped to the location specified by the first match. In addition, whenever "Personal Care — Wash" is specified, the location is mapped to the bathroom, since all personal washing activities are assumed to occur in the bathroom in the modeled house. The criteria used for identifying time periods eligible for each possible water use are given in Table 7.5.

After identifying all eligible periods for the given water use, the mean duration of the water use is subtracted from the end of each eligible time period to prevent the activity from starting too close to the end of that time period. Then all eligible time periods are mapped collectively to a 0 to 1 time scale, shown in Figure 7.2. Subsequently, a Poisson process is used to specify the time between the starting times of successive events by sampling a random number from an exponential distribution with the rate parameter, lambda (λ), equal to the daily frequency. The

TABLE 7.5
Identifying Eligible Water Use Intervals

	Water Source		Eligible Location Codes[a]	Eligible Activity Codes[a]
Source Description	Source Number	Model Location Compartment #		
Shower	1	6	5	40
Bath	2	5	5	40
Toilet	3 & 4	5	5	0
Dishwasher	5	4	0	0
Clothes Washer	6	3	0	0
Faucet (Kitchen)	7	4	3	0
Faucet (Bathroom)	8	5	5	0
Faucet (Laundry)	9	3	9	0

[a] An eligible code of 0 in the water use file matches all codes in the activity pattern.

sampled number will be used to place the starting time of the water-use event by adding it to the start time of the previous event.

This process will be repeated until the next start time falls beyond the end of the last eligible time period. This will result in a simulated frequency of water-use events, on average, equaling the specified frequency.

7.4.2 SIMULATING WATER USE DURATIONS

After the water-use occurrences have been specified, the water-use durations are simulated using the mean and standard deviation of event durations specified in Table 7.4. The toilet is modeled as a standing pool of water with the water-use duration being the variable time between flushes. These duration values are based on information gathered in surveys of water uses (U.S. EPA, 1997). The event duration is assumed to be lognormally distributed, such that:

$$Y \sim Lognormal(\xi, \phi)$$

where: y is the event duration, a lognormally distributed random variable
ξ, ϕ are the parameters of the lognormal distribution

The mean and standard deviation, characteristics of the water-use duration, are converted to the parameters of a lognormal distribution by the following equations:

$$\phi = \left[\ln\left(\frac{\sigma^2}{\mu^2} + 1 \right) \right]^{1/2}$$

$$\xi = \ln(\mu) - \frac{\phi^2}{2}$$

where: μ = the sample mean duration
         σ = the sample standard deviation of the duration

After a random number from the above distribution is simulated, the event duration is assigned this value. If the water-use event continues beyond the end of the occupant's current activity period, the water-use event will be truncated at the end of this activity period. This will happen infrequently, since the eligible time period for simulating the occurrences of the water-uses was shortened by the mean water-use duration, as described above.

7.4.3 HUMIDIFIER WATER USES

Humidifiers are used to evaporate or aerosolize water into the air to increase the relative humidity to improve comfort to the occupants of the home. In the process, any contaminants contained in the water are also emitted into the air. Household tap water is typically used in the humidifier. For this reason, the humidifier may be an important source for some contaminants, particularly lower volatility species. Information in the literature about the typical use pattern for humidifiers is limited. Tsang and Klepeis (U.S. EPA, 1997, Table 15.31) studied the frequency of use in a survey of a total of 1047 persons. The study found that approximately 42% of those persons questioned (135 of 320) were using the humidifier nearly every day during the winter season, 23% (58 of 257) during the spring season, 21% (56 of 269) during the summer season, and 25% (51 of 201) during the fall season. Other survey respondents reported using the humidifier fewer times a week, with a small percentage (less than 4%) reporting never using the humidifier. Information about duration of use was not gathered in this study or found in the literature.

7.5 MODELING CONTAMINANT EMISSIONS DURING HOUSEHOLD WATER USES

Chemicals in the water supply are made available for uptake into the body in vapor, aerosol, and liquid phases. Volatile chemicals are at least partially released into the vapor phase, whereas less volatile chemicals will be found mainly in the liquid (water) and aerosol phases. Uptake can occur by ingestion, dermal, and inhalation routes; only the dermal and inhalation routes are considered in this case study.

Volatile and semivolatile chemicals present in the water supply are released into the air during virtually all household water uses. The household water uses considered in this case study are: showers, faucets, bathtubs, toilets, dishwashers, clothes washers, and humidifiers. The models used to represent the emission processes for each water use are presented below.

7.5.1 VOLATILIZATION MODELS

The volatilization rate is primarily a function of the water temperature, surface area, the concentration difference between the liquid and gas phases, and the chemical's diffusion coefficient and Henry's Law constant. Therefore, the type of flow model will be different for the different water uses. Two types of flow models are used to model the emissions during the household water uses: the plug flow model and the completely mixed flow model (CMFM). Water-using devices do not conform entirely to either of these models. The plug flow model is an idealized representation of the mass transfer from a flowing water supply. The CMFM is an idealized representation of a well-mixed, stationary volume of water. However, both continuous flow water uses and stationary water uses may have several mass-transfer mechanisms occurring simultaneously. For example, when drops are formed in a shower, there are several distinct mass-transfer regions, including drop formation, drop acceleration to terminal velocity, the fall of the drop at terminal velocity, and coalescence on impact at the base of the shower. Similarly, stationary volumes of water have comparable mass-transfer regions during filling and emptying, in addition to the mass-transfer mechanism assumed during the stationary period. The assumed models lump the contributions of the various mass-transfer regions into a single average parameter, $K_{OL}A$, as defined below.

These models and their application to each of the water uses are described below.

7.5.1.1   Emission Rate for the Plug Flow Case

A plug flow model represents water uses where the water is continually renewed, such as the shower and faucets. In these cases the water enters from the source (e.g., a faucet) and leaves at another location (e.g., a drain). The volume of water in contact with the air generally remains the same over the time period of water use, though the water composition is continually renewed at the faucet. The following equation is used to represent volatilization from plug flow sources (Equation 10, Chapter 4, Section 2):

$$C_{out} = C_{in}\exp(Z) + \left(\frac{y}{H}\right)(1 - \exp(Z)) \qquad (7.1)$$

where: C_{in} = concentration in the water supply (mass/volume)
C_{out} = concentration in the water as it flows down the drain (mass/volume)
II   = dimensionless Henry's Law constant
y   = concentration in the air surrounding the water stream (mass/volume)

$$Z = -\frac{K_{OL}A}{Q_L} \qquad (7.2)$$

K_{OL} = overall mass-transfer coefficient (L/time)
Q_L = volumetric flow rate of the water (volume/time)
A = interface area between water and air (L^2)

$$\frac{1}{K_{OL}A} = \frac{1}{K_L A} + \frac{1}{HK_G A}$$

K_L = liquid phase mass-transfer coefficient (L/time)
K_G = gas phase mass-transfer coefficient (L/time)

By manipulating Equation 7.1 and assuming a constant air concentration over a given time step, the following equation is derived:

$$S = K_V \left(C_{in} - \left(\frac{y}{H} \right) \right)$$ (7.3)

where: S = source strength (mass/time)
K_V = volatilization coefficient = $Q_L (1-\exp(Z))$ (7.4)

7.5.1.2 Emission Rate for the CMFM Case

A completely mixed flow model (CMFM) represents water uses that can be best characterized as a system that contains a reservoir of water that remains in contact with the indoor air for a significant period of time. These include the clothes washers, dishwashers, and toilets. The toilet is a special case, where the water remains in contact with the air until the subsequent flush. The following equation is used to represent emissions from these sources (after Equation 16, Chapter 4, Section 2):

$$S = K_{OL} A \left(C - \frac{y}{H} \right) = K_V \left(C - \frac{y}{H} \right)$$ (7.5)

where: K_V = volatilization coefficient = $K_{OL} A$ (7.6)
C = time-varying concentration in the CMFM water (mass/volume)

7.5.2 Model Inputs for Water-Using Devices

Each of the household water-using devices is represented assuming either a plug flow model or a CMFM model. The shower and faucets are modeled as plug flow. The toilet, bathtub, clothes washer, and dishwasher are modeled as CMFMs. The representation used for this case study is reflective of currently available information. While it is understood that each of the CMFM type sources have a plug flow component while they are being filled, this case study neglects the plug flow component because there have been relatively few studies quantifying emission behavior during the filling of these primarily CMFM sources.

To apply these models for estimating the extent of volatilization from household water, estimates of individual mass-transfer coefficients are required for each water-using device at the specified operating condition. Since no published information is available for mass-transfer coefficients in dishwashers, baths, and toilets, an estimate was obtained using the reported transfer efficiencies (TEs) for radon. In addition, an appropriate flow model had to be selected in order to calculate $K_{OL} A$ from TE using Equations 4.2.20 or 4.2.22. Once $K_{OL} A$ was determined, it was assumed equal to $K_L A$ (since radon is highly volatile) and an estimate of $K_G A$ was obtained using

TABLE 7.6
Dimensionless Henry's Law Constant

Temperature (°C)	Chloroform H	Methyl parathion H
20	0.12[a]	8E-6[b]
25	0.15[b]	
30	0.19[c]	
35	0.24[c]	
40	0.30[c]	

[a] Selleck et al. 1988.
[b] Mackay and Shiu, 1981.
[c] H = ((1.3585E8)/T)10^(-1930/T), after Selleck et al. 1988 (Valid for Chloroform only).

a KG/K$_L$ ratio of 40 (assumed to be equal to the average value for showers and faucets). The estimated values of the mass-transfer coefficients for each of the water-using devices are summarized in Chapter 4, along with a more detailed discussion of the assumptions, limitations, and accuracy of these representations. These estimates are preliminary and should only be considered accurate to within an order-of-magnitude.

Two of the compounds being considered in this modeling study, chloroform and methyl parathion, have the ability to enter the gas phase. The third compound, chromium, is a metal and does not enter the gas phase. The Henry's Law constants for these compounds found in the literature are given in Table 7.6. Henry's Law constant for methyl parathion was found for a water temperature of 20°C only. Since this value is very low, relatively small variations for different water temperatures are expected to have a minor impact.

Other model inputs for volatilization used in this case study are given in Tables 7.7 and 7.8 for chloroform and methyl parathion, respectively. The water temperatures for various water uses vary significantly across individuals and households. Hot water heaters can be set to temperatures in excess of 45°C and water typically enters the home at a temperature of less than 15°C. The water temperature for most water uses can be adjusted within this range to suit individual preferences. The individual user also has significant control of the water temperature of dishwashers and clothes washers. Clothes washers are typically constrained to the same range of temperatures as other household water uses, whereas dishwashers often have heating units that can heat the water up to as high as 60°C. Since no survey information was found in the literature on actual water temperatures used in households, the values given in Tables 7.7 and 7.8 were selected to approximate the temperatures believed to be for the given water use.

7.5.3 AEROSOL EMISSIONS

Aerosols are produced and released into the air in the home primarily by showers and humidifiers. Water droplets are subject to removal by evaporation, deposition, and transport out of the home.

TABLE 7.7
Model Parameters for Volatilization of Chloroform and Methyl Parathion

							Chloroform	Methyl Parathion
	Parameters Applied to Both Chloroform and Methyl Parathion							
Water Source	Water Temperature (°C)	Assumed Flow Model[a]	$K_L A$ (L/min)	$K_G A$ (L/min)	$K_{OL} A$ (L/min)	Volatilization Coefficient (L/min)	Henry's Constant H	Henry's Constant H
Shower	40	PFM	15	250	14.2	7.6	0.3	8E-6
Bath	40	CMFM	7	280	6.8	6.8	0.3	8E-6
Toilet	25	CMFM	0.07	2.8	0.07	0.1	0.15	8E-6
Dishwasher	40	CMFM	9	360	8.8	8.8	0.3	8E-6
Clothes Washer	35	CMFM	1	40	1.0	1.0	0.24	8E-6
Faucet (Kitchen)	35	PFM	1.5	100	1.48	1.29	0.24	8E-6
Faucet (Bathroom)	35	PFM	1.5	100	1.48	1.29	0.24	8E-6
Faucet (Laundry)	30	PFM	1.5	100	1.48	1.29	0.19	8E-6

[a] PFM refers to plug flow model, and CMFM refers to completely mixed flow model.
K_L = Liquid-phase mass-transfer coefficient.
K_G = Gas-phase mass-transfer coefficient.
K_{OL} = Overall mass-transfer coefficient.

TABLE 7.8
Distribution of Aerosol Mass in Showers
Based on the Distribution Shown in Figure 7.4

Range, µm	Fraction of Mass
0–2.5	0.193
>2.5	0.807

Evaporative processes remove the water from the droplet, leaving the nonvolatile portion as a particle. The time for evaporation of the water from a particle is dependent upon the mass of water and the relative humidity in the surrounding air. According to Hinds (Hinds 1982), a water particle in air at 50% relative humidity with an initial diameter of 10 µm will fully evaporate in less than 1 second, while a particle with an initial diameter of 50 µm will evaporate in approximately 4 seconds.

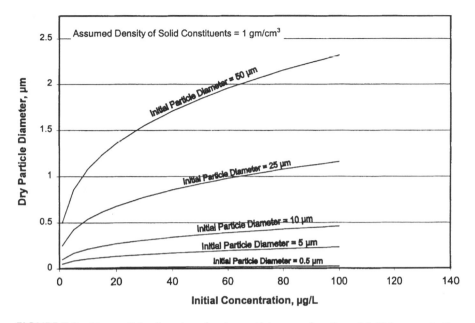

FIGURE 7.3 Dry particle diameter of water particles as a function of initial concentration and initial particle diameter.

In the case of volatile compounds like chloroform, the compound will volatilize before the water evaporates. However, in the case of chromium and methyl parathion, it can be assumed that after the water evaporates, the contaminant remains. Figure 7.3 presents a family of curves for various initial drop sizes, which show the relationship between concentration in the water supply and dry particle diameter. These curves are developed on the assumption that the density of the solid portion is 1 gm/cm^3. Actual and apparent particle density varies over the range from less than 1 to greater than 5, depending on the density of their parent material and the amount of void space within an agglomerated structure. The chosen value of 1 gm/cm^3 can be viewed as a relatively low value, resulting in particles of larger diameters for a given concentration. Using this assumed dry density, water supplies with a relatively high initial concentration generate particles with dry diameters of less than 2.5 μm. Therefore, this case study assumes that dry particle diameters for aerosols generated during the use of a water supply are uniformly distributed between 0 and 2.5 μm, with the exception of emissions in the vicinity of the shower (see Section 7.5.3.1).

Deposition is a function of the droplet size and the turbulence of the surrounding air. According to Hinds (1982), the settling velocity of 10 and 50 μm particles is 0.3 cm/s and 7.5 cm/s, respectively. This corresponds to a half-life with respect to deposition of approximately 11 minutes and 28 seconds, respectively. In both cases, the majority of the drops would evaporate long before they are deposited to a surface. For these reasons, deposition is ignored in this case study except in the shower, where evaporation is slower due to the high relative humidity, and deposition is enhanced due to the close proximity of surfaces.

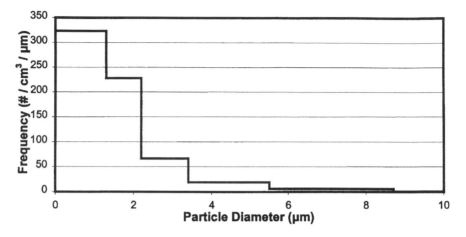

FIGURE 7.4 Size distribution of aerosols produced in showers (based on Keating and McKone, 1993).

7.5.3.1 Modeling of Aerosol Emissions in Showers

The distribution of aerosols produced in the shower is a function of the showerhead and the water flowrate. Relatively few researchers have characterized the distribution of aerosols in showers. A study by Keating and McKone (1993) characterized the aerosol size distribution present in the shower while in use. Using the information provided by Keating and McKone and several additional assumptions, the size distribution shown in Figure 7.4 approximates aerosol concentrations in showers. The average case for the showers in the study is estimated as the mean of the maximum and minimum aerosol concentrations.

   Integration of the distribution in Figure 7.4 for mass in the respirable (0 to 2.5 µm) range yields the values in Table 7.8. For the distribution shown in Figure 7.4, 19% of the mass is in the respirable range with the remaining 81% in the 2.5 to 10 µm range.

   In another study, Gunderson and Witham (1988) found that the aerosol production during use of the shower could be approximated using a simple linear model, given by the following equation:

$$C_a\left(\mu g\ m^{-3}\right) = A\ C_w\left(mg\ L^{-1}\right) \tag{7.7}$$

where: A = an empirical constant found to vary between 0.03 and 0.2. The exposure due to aerosols originating in the shower is expected to be minor compared to other routes. For this reason, a value of 0.2 is used in this case study to represent a worst case scenario.

C_w = the chemical concentration in the aqueous phase.

C_a = the chemical concentration in the aerosol phase (µg compound in aerosol phase/m³ air)

Equation 7.7 gives the total airborne concentration of the contaminant in the aerosol phase while the shower is being used. Equation 7.7 can be modified to allow the same units for C_a and C_w:

$$C_a = a \, C_w \tag{7.8}$$

where: $a = A * 10^{-6} \approx 2 \times 10^{-7}$

The parameter "a" in Equation 7.8 represents the volumetric aerosol concentration. By making several assumptions, the predictions of this equation can be compared with the results presented by Keating and McKone for an evaluation of consistency. The distribution presented in Figure 7.4 is assumed and integrated for liquid mass, with the following results:

Aerosol concentration $\approx 4 \times 10^{-9}$ m^3 H$_2$O/m^3 air

The value reported by Gunderson and Witham for A of between 0.03 and 0.2 corresponds to a range for the parameter a as follows:

3×10^{-8} m^3 H$_2$O/m^3 air $< a < 2 \times 10^{-7}$ m^3 H$_2$O/m^3 air

Equation 7.7 predicts an aerosol concentration approximately 10 to 100 times larger than that measured by Keating and McKone. Given the variability in showerheads, water pressure, water flow rates, bathroom and shower volumes, and other factors influencing aerosol production rates and concentrations, this is deemed to be acceptable for the purposes of this case study. A value of 2×10^{-7} m^3 H$_2$O/m^3 air is used in this case study to represent the higher end of the expected aerosol concentrations, since aerosols are not a major contributor to exposure.

Equation 7.7 is representative of the concentration of nonvolatile species in the aerosol phase. The aerosols are transported throughout the home through interzonal flows and removed from the home by exfiltration.

The total aerosol chemical concentration in the bathroom and shower is set to C_a when the shower is being used, and assumed to follow the size distribution shown in Figure 7.4. Upon leaving the shower and bathroom area, the aerosols are then assumed to become dry particles due to rapid evaporation, as discussed above.

7.5.3.2   Modeling of Aerosol Emissions in Humidifiers

The humidifier modeled in this case study is assumed to be an ultrasonic humidifier that emits aerosols at a constant rate. Volatile compounds are assumed to instantaneously volatilize upon aerosolization. Semi- and nonvolatile compounds are assumed to remain entirely in the aerosol phase.

Humidifiers are very efficient at creating particles < 2.5 μm. Because these particles are in the range for maximum deposition in the deep lung, they are expected to be an important contributor to exposure. The overall humidifier emission rate can be approximated as (Highsmith et al., 1988, 1992):

$$E = C_w \, \text{WCR} \, f \tag{7.9}$$

where: E = emission rate, mg/hr
        C_w = concentration in water, mg/L
        WCR = water consumption rate = 0.5 L hr^{-1}
        f = 0.9 for particles < 2.5 μm
        f = 0.1 for particles between 2.5 μm and 10 μm

As shown earlier, the water content in these particles can be expected to evaporate in a few seconds. The resulting dry particle size of any nonvolatile constituents can be expected to be less than 2.5 μm. For this reason, particles formed at the emission rate given by Equation 7.9 are assumed to be uniformly distributed between 0 and 2.5 μm. The aerosols are then assumed to be distributed throughout the house and removed to the outdoors through normal transport processes.

7.6   MODELING ABSORBED DOSE

The term *dose* has many meanings, depending on the context. This case study will estimate the dose as defined as (U.S. EPA, 1997):

> The absorbed dose is the amount crossing a specific absorption barrier (e.g., the exchange boundaries of skin, lung, and digestive tract) through uptake processes.

The absorbed dose estimated in this case study is the result of the chemical crossing the skin boundary (dermal) and the lung boundary (inhalation). In addition, the potential dose is also estimated for these boundaries, defined as (U.S. EPA, 1997):

> The potential dose is the amount ingested, inhaled, or applied to the skin.

The procedures used to calculate the absorbed dose are described below.

7.6.1   Absorbed Dose due to Vapors

In the lungs, the transfer of an inhaled compound from the vapor phase into the bloodstream is modeled as a system that achieves equilibrium instantaneously, as shown in Figure 7.5. At equilibrium, Henry's Law constant describes the partitioning of the chemicals between the vapor and liquid (blood) phases. The chemical uptake into the blood estimated by this technique is consistent with recent studies (Johanson, 1991) showing that approximately 60 to 80% of inhaled polar compounds absorb into the bloodstream. This technique offers a good approximation of uptake for short exposures (e.g., showers of 7 to 10 minutes); however, for longer exposures it likely overpredicts because it does not account for chemical buildup in the bloodstream, metabolism rate, excretion, etc. To more accurately describe the lung–blood exchange, a full physiologically based pharmacokinetic (PBPK) model should be linked to this exposure model. The following steps are used to approximate chemical uptake into the blood:

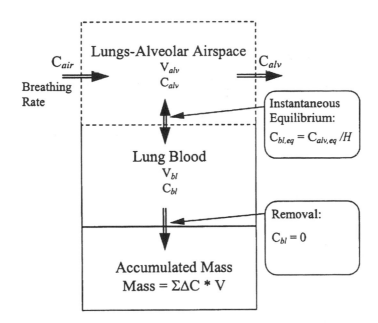

FIGURE 7.5   Model representation of uptake by the inhalation route.

The equilibrium concentrations in the alveolar blood and the alveolar air are as follows:

$$C_{bl,eq} = C_{alv,eq}/H \qquad (7.10)$$

where: $C_{bl,\ eq}$ = concentration in the lung blood after equilibrium is reached
with alveolar air
$C_{alv,eq}$ = concentration in the alveolar air after equilibrium is reached
with lung blood
H   = dimensionless Henry's Law constant

The equilibrium concentration in the lung blood and alveolar air is calculated by the following equations:

$$C_{bl,eq} = \left(C_{bi}V_{bl} + C_{air}V_{alv}\right)/\left(V_{bl} + H\,V_{alv}\right) \qquad (7.11)$$

and

$$C_{alv,eq} = H\,C_{bl,eq} \qquad (7.12)$$

where: V_{alv} = alveolar volume for a time step = breathing rate multiplied by
time step
V_{bl} = lung blood volume for a time step = cardiac output multiplied by
time step

TABLE 7.9
Deposition Data for the Human Respiratory System

Particle Diameter	Fraction of Inspired Mass Deposited			
(μm)	Nose	Airways	Alveolar	Total
0.06	0	0.073	0.46	0.53
0.2	0	0.025	0.20	0.22
1	0.27	0.016	0.14	0.42
2	0.59	0.031	0.20	0.82
3	0.71	0.043	0.19	0.94
6	0.92	0.048	0.05	1.0
10	1	0	0	1.0

C_{bi} = bulk blood concentration at the start of the time step
C_{air} = air concentration entering lungs
Cardiac output ≈ 5 L hr^{-1} kg^{-1}, for a normal adult (Berne and Levy, 1993)
Body weight ≈ 72 kg
Breathing rate ≈ 20 m^3 day^{-1}, for a normal adult (U.S. EPA, 1997)

The mass accumulated in the lung blood is assumed to accumulate in the body, and the lung blood concentration is reset to zero between time steps. The cumulative mass accumulated in the body is calculated as follows:

$$\text{Mass absorbed} = \sum \Delta C * V \quad \text{over all time steps} \tag{7.13}$$

The air concentration outside the body is assumed to be unaffected by the mass transferred into the blood.

7.6.2 ABSORBED DOSE DUE TO AEROSOLS

The human upper respiratory system filters essentially 100% of aerosols with a diameter greater than approximately 7 μm. However, particles less than 7 μm are not filtered as efficiently, and therefore a significant fraction of the inhaled mass reaches the alveolar region. The fraction of the inhaled aerosol mass deposited in each region (nose, airways, and alveolar) is dependent on the particle size distribution, upstream filtering, and the respiration rate. A large mass fraction of the particles less than 2.5 μm reach the alveolar region. Table 7.9 presents deposition data for various regions of the human respiratory system (Yeh and Schum, 1980).

7.6.2.1 Absorbed Dose in Shower Area due to Aerosols

To calculate the absorbed chemical dose in the shower area due to aerosols, Figure 7.4 is assumed to represent the underlying particle-size distribution emitted during shower use. This distribution is integrated over the appropriate size intervals

TABLE 7.10

Aerosol Deposition in the Human Respiratory System for the Particle Distribution Shown in Figure 7.4

Location	Fraction of Inspired Mass in Size Range Deposited	
	0–2.5 μm	>2.5 μm
Nose	0.490	0.898
Airways	0.028	0.019
Alveolar Region	0.181	0.066

TABLE 7.11

Aerosol Deposition in the Human Respiratory System for a Uniform Particle Distribution Between 0 and 2.5 μm

Location	Fraction of Inspired Mass Deposited
Nose	0.540
Airways	0.030
Alveolar Region	0.184

to estimate the fraction of the total mass in each interval. This estimated mass is then combined with the deposition data appropriate to the size interval to estimate the fraction of the aerosol mass deposited in each region of the human respiratory system. The resulting estimated fractional masses are shown in Table 7.10.

7.6.2.2 Absorbed Dose Outside of Shower Area due to Aerosols

In all areas other than the bathroom, a uniform distribution between 0 and 2.5 μm particle diameters is assumed to be the underlying particle-size distribution. This distribution is integrated over the intervals and combined with the deposition data appropriate to the size interval to estimate the fraction of the aerosol mass deposited in each region of the human respiratory system. The resulting estimated fractional masses are shown in Table 7.11 as a fraction of the total available mass.

7.6.3 ABSORBED DOSE DUE TO DERMAL CONTACT

The concentration of the chemical in the water in contact with the skin is assumed to be the same as the concentration in the water supply. The water concentration at the time of contact with the skin is dependent on the amount of the chemical that has volatilized prior to contact. For more volatile species, a significant amount of the chemical may have volatilized out of the water prior to water use. However, in modeling dermal exposure, this case study does not attempt to consider volatilization, resulting in a worst-case scenario for dermal uptake estimation.

The absorbed dose through the dermal route is dependent on many factors, including the concentration in the water contacting the skin, the skin contact location and area, the length of contact, water temperature, and chemical properties. Dermal contact with the household water supply occurs primarily during two types of water-using activities: (1) showering and bathing and (2) faucet use. This case study assumes that for any bathing activity, the entire body is in contact with the contaminated water and for any faucet use only the hands are in contact. The body surface areas corresponding to these assumptions are estimated to be 18,000 cm² for the entire body and 720 cm² for both hands (U.S. EPA, 1997). The contact is assumed to occur for the time period that the water use is active. The water in contact with the skin is assumed to have the initial water concentration. The water temperature has been shown to impact the rate of uptake for chloroform (Gordon et al., 1997), and is likely to impact the uptake rate for other chemicals as well.

Two alternative models are used in this case study to estimate dermal uptake. The predictions of the models will be evaluated and compared.

7.6.3.1 Traditional (Steady-State) Approach to Dermal Absorption

Traditional models assume steady-state transfer across the skin and estimate the dermal penetration as a linear function of the time of exposure, as presented in Chapter 6. The governing equation is as follows (Chapter 6, Section 6.2.1, Equation 6.4):

$$M = A \, P_m \, C_w^o t_{exp} \tag{7.14}$$

where: M = mass crossing the skin
A = area of skin in contact with water
P_m = permeability coefficient representing the chemical's ability to cross the skin membrane $\approx P_{sc}$
P_{sc} = permeability coefficient of the stratum corneum
C_w^o = concentration of solute in the aqueous solution
t_{exp} = duration of exposure

Refer to Chapter 6, Section 6.2.1 for further description of the model parameters and discussion on advantages and disadvantages of using this model. The permeability coefficients used for each of the chemicals modeled in this case study are given in Table 7.12. The water concentration in contact with the skin is assumed to be the same as the concentration in the water supply, as discussed above.

7.6.3.2 Membrane Approach to Dermal Absorption

Other investigators have used dynamic models for the flux of the chemical across the skin, as presented in Chapter 6, Section 6.4.1. This case study implements the technique presented by Cleek and Bunge (1993) as given by the following equations:

TABLE 7.12
Parameters for Skin Penetration Models

Chemical	P_{sc} (cm/hr)	log $K_{o/w}$	$R_{sc/w}$	MW	t_{lag}^g (hrs)
Chloroform	0.160[a]	1.97	25[e]	119.38	0.042
Methyl parathion	0.005[b]	2.86	107[e]	263.2	5.7
Chromium	0.0021[c]	N.A.[d]	0.1[f]	52.0	0.013

[a] Nakai, J. et al., 1985.
[b] Calculated using Equation 7.19.
[c] Baranowska-Dutkiewicz, B., 1981.
[d] This value not available in literature.
[e] Calculated using Equation 7.17.
[f] Estimate based on measurements for ionized chemicals summarized in Vecchia, B.E., 1997.
[g] Calculated using the definition of t_{lag} following Equation 7.16.

$$M_{in} = AC_w^o \sqrt{\frac{4R_{sc/w}L_{sc}P_{sc}t_{exp}}{\pi}} \qquad \text{when } t_{exp} < 2.4t_{lag} \qquad (7.15)$$

and

$$M_{in} = AC_w^o \left(P_{sc}t_{exp} + \frac{R_{sc/w}L_{sc}}{3} \right) \qquad \text{when } t_{exp} \geq 2.4t_{lag} \qquad (7.16)$$

where:
M_{in} = mass entering the skin
A = area of skin in contact with water
C_w^o = concentration of solute in the aqueous solution
t_{exp} = duration of exposure
$R_{sc/w}$ = equilibrium partition coefficient stratum corneum (sc) and water (w)
L_{sc} = diffusion path length through the stratum corneum (sc)
P_{sc} = permeability coefficient of the stratum corneum (sc)
t_{lag} = $\dfrac{R_{sc/w}L_{sc}}{6P_{sc}}$ = estimated time to reach steady state

When better information is not available, the values for $R_{sc/w}$, L_{sc}, and P_{sc} can be estimated as follows:

$$\log R_{sc/w}[\text{unitless}] = 0.71 \log K_{o/w} \qquad (7.17)$$

$$L_{sc} \approx 16\,\mu\text{m} = 1.6 \times 10^{-3}\,\text{cm} \qquad (7.18)$$

$$\log P_{sc}[\text{cm/hr}] \cong \log P_m = -2.72 + 0.71\,\log K_{o/w} - 0.0061\ MW \qquad (7.19)$$

Refer to Chapter 6, Section 6.4.1 for further description of the model parameters and discussion on advantages and disadvantages of using this model. The parameters used for each chemical modeled in this case study are given in Table 7.12. The water concentration in contact with the skin is assumed to be the same as the concentration in the water supply, as discussed above.

7.7   RESULTS

The exposures and dose resulting from the above-described scenario are estimated assuming the water supply is contaminated with each of the compounds (chloroform, methyl parathion, and chromium) at a concentration of 1 mg/L. Each simulation follows the modeling techniques described in the above sections. The simulations are repeated for 1000 iterations, each iteration sampling the activity pattern database for two adults (1 female and 1 male, each between the ages of 18 and 34) with the assumption that these two adults share a household. The results of the modeling runs are presented in the following forms: results of one selected case (1 case of the 1000 iterations); the results of a simulation where a humidifier is used; and the collective results for the entire population group (2000 individuals). Additionally, the collective population group results are separated into two groups: those who showered or bathed and those who did not.

7.7.1   RESULTS OF A SELECTED CASE

To illustrate specific behavior of the model, the results of one typical case are presented. The sampled activity patterns are shown in Table 7.13 for Adult 1 (female, age 18 to 34) and Adult 2 (male, age 18 to 34). The simulated water uses for this case are given in Table 7.14.

For these water uses and activity patterns, the resultant air concentrations and personal uptakes of the chemicals (chloroform, methyl parathion, and chromium) are modeled and presented in Figures 7.6, 7.7, and 7.8, respectively. Table 7.15 summarizes the dermal dose for each individual and chemical, grouped by the three exposure mediums estimated in this study: gas, aerosol, and liquid.

The top portion of Figure 7.6 shows the predicted 24-hour time-varying air concentrations for chloroform, while the bottom portion presents the cumulative absorbed dose for the inhalation route. Since it is assumed that volatilization occurs instantaneously from the aerosol phase, no exposure occurs due to the aerosol phase.

Similarly, for the methyl parathion simulation, the 24-hour time-varying air concentrations, the 24-hour time-varying concentration in the aerosol phase, and the cumulative absorbed dose for the inhaled vapor and inhaled aerosol routes are shown respectively in the top, middle, and lower portions of Figure 7.7. Finally, for the

TABLE 7.13
Sampled Activity Pattern

	Adult 1		Adult 2
Location	Enter Time (Hours after Midnight)	Location	Enter Time (Hours after Midnight)
Bedroom	0	Bedroom	0
Living Room	6.5	Living Room	7
Kitchen	7.25	Kitchen	9
Hall/Entry	7.75	Outdoors	9.5
Bathroom	8	Hall/Entry	11
Hall/Entry	8.5	Outdoors	11.25
Outdoors	8.75	Kitchen	16.5
Kitchen	18.08	Outdoors	16.83
Outdoors	18.33	Bathroom	17.5
Kitchen	20.75	Shower	17.8
Living Room	21.25	Bathroom	17.89
Bedroom	21.5	Living Room	18
		Kitchen	20.5
		Bedroom	24

chromium simulation, the 24-hour time-varying concentration in the aerosol phase and cumulative absorbed dose for the aerosol route are shown respectively in the upper and lower portions of Figure 7.8. The air concentrations and inhalation dose are not shown since chromium is a nonvolatile compound and its concentration is negligible in the gas phase.

The dermal dose is estimated by two fundamentally different approaches. The steady-state approach assumes instantaneous steady-state conditions upon contact with the water. It further assumes that the skin has no storage capacity, and therefore no uptake occurs after contact with the water ends. In contrast, the membrane approach assumes a period of non-steady-state flux, where the flux rate gradually increases as it approaches steady state, based on Fick's law. It also assumes that the skin has a storage capacity, and therefore uptake continues after contact with the water ends. Neither approach accounts for volatilization from the skin after exposure, which is important for volatile chemicals such as chloroform.

The dermal results (Table 7.15) show that the total absorbed dose estimated by the steady-state approach (Equation 7.14) is generally lower than that estimated by the membrane approach (Equations 7.15 and 7.16). The dermal exposures for these individuals were shorter than the estimated time required to reach steady state $(2.4*t_{lag})$, resulting in the use of Equation 7.15 for the membrane approach. The estimated dermal doses for chloroform from the steady-state calculations are 34% and 63% of the estimates from the membrane equation for Adult 1 and Adult 2, respectively. For methyl parathion, the estimated dermal doses from the steady-state calculations are only 3 to 4% of that for the membrane equation. Methyl parathion has a lower permeability in skin than chloroform and is more lipophilic and therefore

TABLE 7.14
Simulated Water Uses

Water Source	Start Time (Hours after Midnight)	End Time (Hours after Midnight)	User
Shower	17.74	17.89	Adult 2
Bath	8	8.06	Adult 1
Bath	17.68	17.81	Adult 2
Toilet	3	—	—[a]
Toilet	4.82	—	—[a]
Toilet	8.1	—	—[a]
Toilet	10.82	—	—[a]
Toilet	16.49	—	—[a]
Toilet	18.17	—	—[a]
Toilet	22.26	—	—[a]
Dishwasher	5.3	6.6	Adult 1[b]
Dishwasher	22.24	23.32	Adult 1[b]
Clothes Washer	12.03	12.82	Adult 1[b]
Kitchen Faucet	16.738	16.742	Adult 2
Kitchen Faucet	16.789	16.795	Adult 2
Kitchen Faucet	18.253	18.258	Adult 1
Bathroom Faucet	8.052	8.054	Adult 1
Bathroom Faucet	17.517	17.519	Adult 2
Bathroom Faucet	17.524	17.527	Adult 2
Bathroom Faucet	17.691	17.696	Adult 2
Bathroom Faucet	17.747	17.786	Adult 2
Bathroom Faucet	17.844	17.846	Adult 2
Bathroom Faucet	17.852	17.856	Adult 2
Laundry Faucet	11.108	11.117	Adult 2
Laundry Faucet	11.128	11.136	Adult 2
Laundry Faucet	11.143	11.148	Adult 2

[a] Use of the toilet is assumed to be a property of the house, and not affiliated directly with an occupant

[b] Use of the indicated sources does not require the user's presence

has a longer lag time. Consequently, the membrane approach is likely to have a larger estimate than the steady-state approach because of the quantity stored in the skin. For chromium, both approaches estimate similar, but very small, dermal doses.

7.7.1.1 Results of Humidifier Simulation

Because of the lack of information about typical use patterns for humidifiers, the impact of the humidifier is simulated independent of the other water uses in the home (see Sections 7.4.3 and 7.5.3.2). This simulation is analyzed to assess the relative importance of the humidifier for each of the three contaminants.

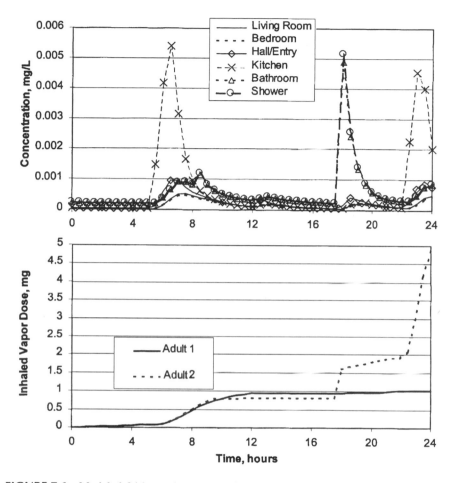

FIGURE 7.6 Modeled 24-hour air concentrations and inhaled vapor dose for chloroform. (Note: Exposure to the chloroform in the aerosol phase is not considered.)

In modeling humidifier emissions, volatilization is neglected because the water mass available for volatilization in a humidifier is very small (less than 1%) compared to other water sources in the home and therefore would have a negligible impact on air concentrations. Since it is assumed that no volatilization occurs, the results of methyl parathion and chromium are identical for the aerosol phase, with the results shown in Figure 7.9. The upper portion of the figure shows the 24-hour time-varying concentration in the aerosol phase and the lower portion shows the cumulative absorbed dose.

7.7.2 RESULTS OF POPULATION DOSE DISTRIBUTION

The results of the 1000 simulations (2000 adults) are analyzed to estimate the relative impact of the various routes of exposure, the distribution of the doses within the population group, and the overall magnitude of the exposure to each compound.

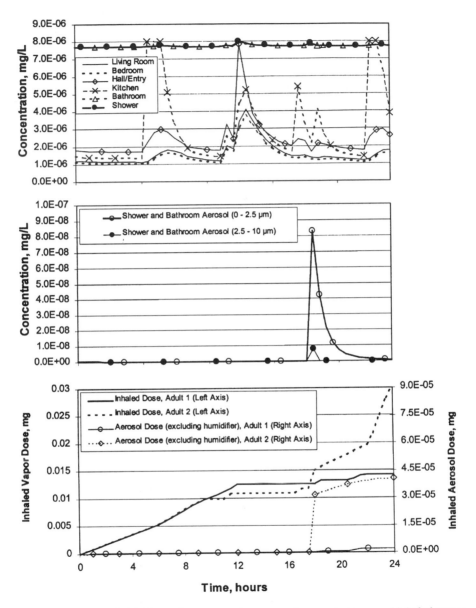

FIGURE 7.7 Modeled 24-hour air concentrations, aerosol concentrations, and inhaled vapor dose for methyl parathion. (Note: These results exclude the humidifier contribution.)

Bathing and showering are recognized for their large impact on exposure, and therefore the results are analyzed and presented separately for the portion of the population that bathed and showered and for the portion that did not. The shower frequency for the simulations was specified as 0.7 showers per day per person, reflecting survey results that 70% of the population takes a shower on any given

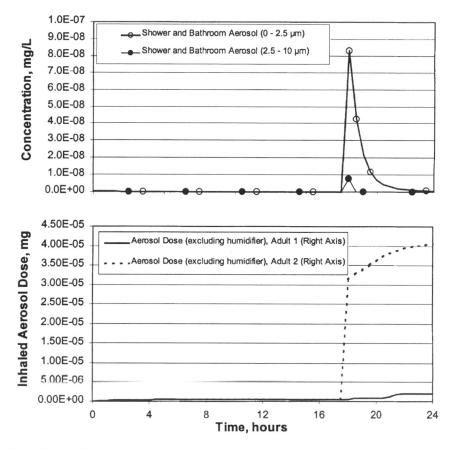

FIGURE 7.8 Modeled 24-hour aerosol concentrations and inhaled aerosol dose for chromium. (Note: These results exclude the humidifier contribution. Chromium is nonvolatile, therefore air concentrations are not considered.)

TABLE 7.15
Modeled 24 Hour Dermal Dose for Chloroform, Methyl Parathion, and Chromium

	Estimated 24 Hour Dermal Dose (mg/day)					
	Chloroform		Methyl Parathion		Chromium	
Uptake Model	Adult 1	Adult 2	Adult 1	Adult 2	Adult 1	Adult 2
Equation 14	0.11	0.77	0.004	0.02	0.0015	0.010
Equations 15 & 16	0.32	1.23	0.12	0.46	0.0024	0.0090

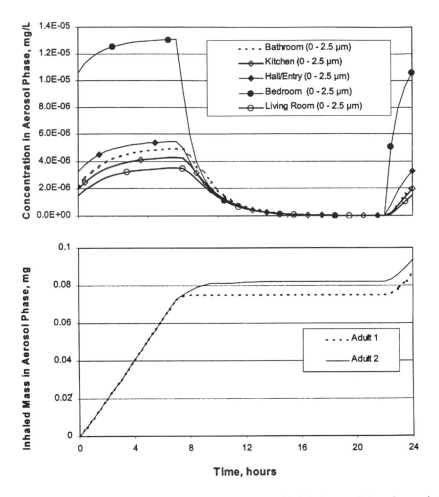

FIGURE 7.9 Modeled 24-hour aerosol concentrations and inhaled aerosol dose for methyl parathion and chromium due to the humidifier.

day. In this case study, however, the simulated frequency was less than 0.4 showers per day. The cause of this poor representation has been traced to inadequacies in the activity pattern database used to represent the population's activities. Only approximately half of the activity patterns recorded in the database allowed a sufficient interval for a shower or bath to potentially occur (i.e., more than 6 consecutive minutes spent in the bathroom).

The absorbed dose through each of the simulated mechanisms (inhaled vapor phase, dermal, and inhaled aerosol phase) is summed to calculate the total absorbed 24-hour dose. Figures 7.10, 7.11, and 7.12 present the distribution of total absorbed dose for chloroform, methyl parathion, and chromium, respectively.

Each of the three simulated uptake mechanisms (inhaled vapor phase, dermal, and inhaled aerosol) is presented below. These results are analyzed and discussed in Sections 7.7.3 and 7.8.

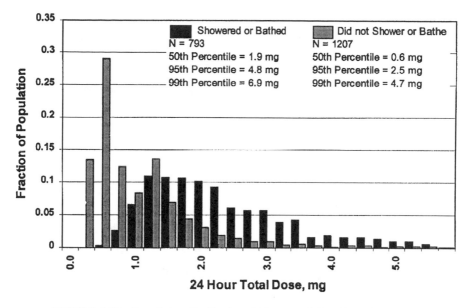

FIGURE 7.10 Population distribution: 24-hour total dose for chloroform.

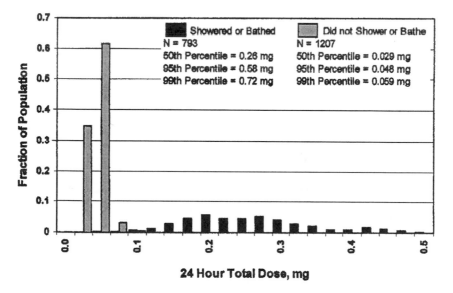

FIGURE 7.11 Population distribution: 24-hour total dose for methyl parathion.

7.7.2.1 Vapor Phase Exposure

The inhaled vapor dose is analyzed for the volatile and semivolatile compounds (chloroform and methyl parathion). Since volatilization of chromium is assumed to be negligible, it is not analyzed for vapor phase exposure. Histograms of the simulated

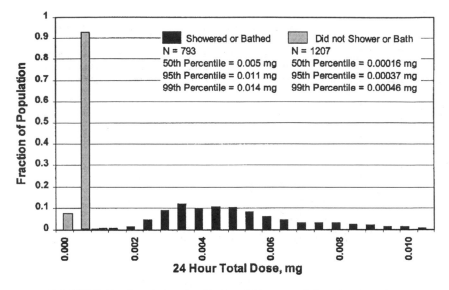

FIGURE 7.12   Population distribution: 24-hour total dose for chromium.

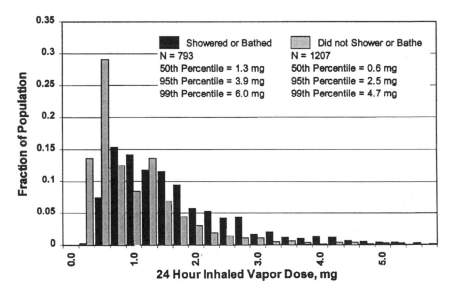

FIGURE 7.13   Population distribution: 24-hour inhaled vapor dose for chloroform.

population doses for the 24-hour potential inhaled vapor dose are presented for chloroform in Figure 7.13 and for methyl parathion in Figure 7.14.

7.7.2.2   Dermal Exposure

Compounds present in the liquid phase enter the body through the skin, resulting in a dermal dose. This case study assumes no volatilization occurs prior to skin contact

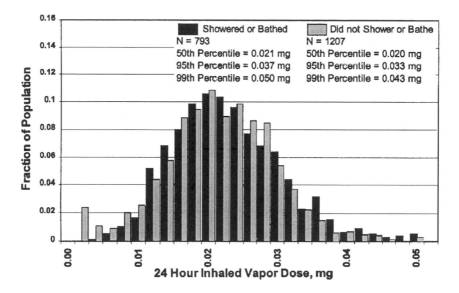

FIGURE 7.14   Population distribution: 24-hour inhaled vapor dose for methyl parathion.

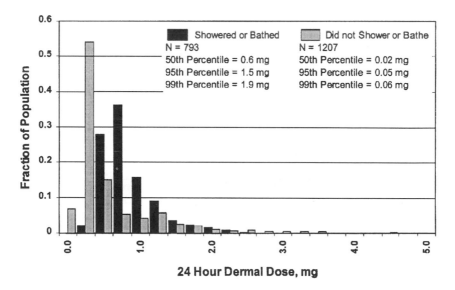

FIGURE 7.15   Population distribution: 24-hour dermal dose for chloroform.

and therefore contact with the liquid phase results in dermal exposure to each of the three contaminants considered in this study.

Figures 7.15, 7.16, and 7.17 present the histograms of absorbed dermal mass (calculated by the membrane approach) for chloroform, methyl parathion, and chromium, respectively.

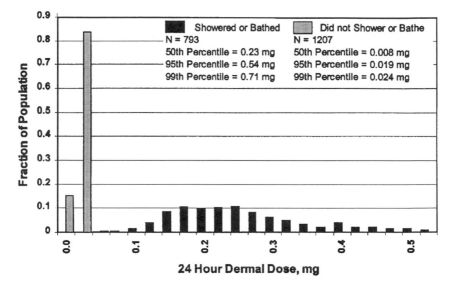

FIGURE 7.16   Population distribution: 24-hour dermal dose for methyl parathion.

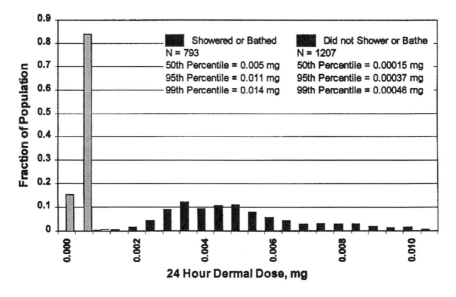

FIGURE 7.17   Population distribution: 24-hour dermal dose for chromium.

7.7.2.3   Aerosol Phase Exposure

The aerosol dose is analyzed for the semivolatile and nonvolatile compounds (methyl parathion and chromium, respectively). Since the volatilization is assumed to be negligible, the mechanism for aerosol exposure is the same for both compounds. The histogram of the simulated population statistics for the 24-hour total aerosol dose in the respiratory tract is presented in Figure 7.18 for methyl parathion and chromium.

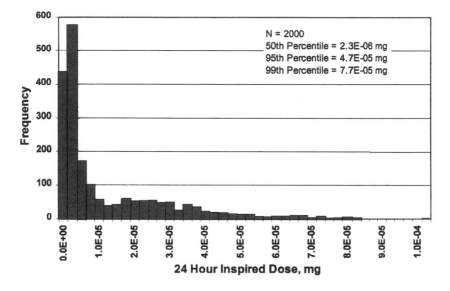

FIGURE 7.18 Population distribution: inhaled aerosol 24-hour dose for methyl parathion and chromium. (Note: These results exclude the humidifier contribution.)

7.7.2.4 Analysis of Aerosol Deposition Location in the Respiratory System

Aerosol phase contaminants are deposited in the respiratory system at various locations (nose, airways, and alveolar) as a function of the particle size. Based on the assumed particle distributions discussed in Sections 7.5.3.1 and 7.5.3.2 and the respiratory system deposition characteristics presented in Section 7.6.2, an analysis of the location of the deposited aerosols is performed for the 2000 adults simulated in this case study. Figure 7.19 presents the fraction the dose deposited in each region of the respiratory system as a function of total 24-hour inhaled aerosol dose. Table 7.16 presents the fraction deposited in each region for the entire population.

7.7.3 DISCUSSION

The three chemicals evaluated in this case study were chosen to represent a range of properties that impact potential inhalation and dermal exposure. These properties include volatility, molecular weight, Henry's Law constant, and skin permeability. As a result, the simulated distributions of absorbed dose varied significantly both in shape and magnitude (see Table 7.17 and Figures 7.10, 7.11, and 7.12).

It is hypothesized that the differing shapes of the distributions seen in Figures 7.10 to 7.17 are the result of the three chemicals' characteristics and transport mechanisms. The population distributions of inhaled vapor doses tend toward lognormal for the high-volatility compound (chloroform) and tend toward normal for the low-volatility compound (methyl parathion). It was assumed that no inhalation exposure to chromium occurred because of its extremely low volatility.

In the case of chloroform, its high volatility leads to its rapid movement from liquid to air. Large water-use sources, such as showers, become dominant sources

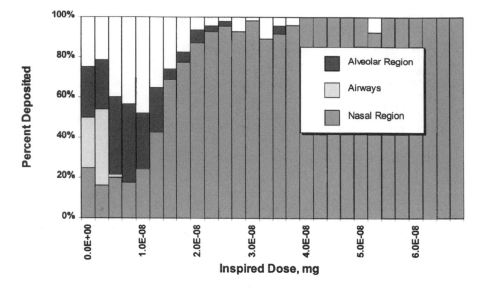

FIGURE 7.19 Population distribution: location of deposition in the respiratory tract as a function of 24-hour inhaled aerosols mass. (Note: These results exclude the humidifier contribution.)

TABLE 7.16
Fraction of Total Inhaled Aerosol Mass
Deposited in the Various Respiratory Regions

Region	Fraction Deposited
Nose	0.660
Airways	0.024
Alveolar	0.133
TOTAL	0.817

---

with respect to exposure because a large fraction of the total contaminant mass in the water is ultimately released to the air. As a result, concentrations change rapidly in the vicinity of the source, and factors such as the length of the shower and the location of the individual with respect to the water use have an extremely large impact on the dose. These factors tend to be lognormally distributed over the population, yielding exposure distributions tending toward lognormal. This effect is further noticed in the comparison of the distributions of doses for those who showered or bathed with those who did not, as seen in Figure 7.13.

In the case of methyl parathion, the low volatility leads to very slow movement from the liquid phase to the air. Therefore, sources with long contact time, such as toilets, dishwashers, and clothes washers, become the important sources of airborne concentrations and provide the dominant inhaled vapor exposure. As a result, the

TABLE 7.17
Summary of Population Exposure Distributions Based on Unit Concentrations

| Chemical | Exposure Route | Population Distribution: 24 Hour Dose (mg) | | |
		50th Percentile	95th Percentile	99th Percentile
Chloroform	Occupants who showered or bathed			
	Inhalation	1.3	3.9	6.0
	Aerosol[a]	0	0	0
	Dermal	0.6	1.5	1.9
	Total	1.9	4.8	6.9
	Occupants who did not shower or bathe			
	Inhalation	0.6	2.5	4.7
	Aerosol[a]	0	0	0
	Dermal	0.02	0.05	0.06
	Total	0.6	2.5	4.7
Methyl Parathion	Occupants who showered or bathed			
	Inhalation	0.021	0.037	0.050
	Aerosol[a]	2.3E-06	4.7E-05	7.7E-05
	Dermal	0.23	0.54	0.71
	Total	0.26	0.58	0.72
	Occupants who did not shower or bathe			
	Inhalation	0.020	0.033	0.043
	Aerosol[a]	2.3E-06	4.7E-05	7.7E-05
	Dermal	0.008	0.019	0.024
	Total	0.029	0.048	0.059
Chromium	Occupants who showered or bathed			
	Inhalation	0	0	0
	Aerosol[a]	2.3E-06	4.7E-05	7.7E-05
	Dermal	0.005	0.011	0.014
	Total	0.005	0.011	0.014
	Occupants who did not shower or bathe			
	Inhalation	0	0	0
	Aerosol[a]	2.3E-06	4.7E-05	7.7E-05
	Dermal	0.00015	0.00037	0.00046
	Total	0.00016	0.00037	0.00046

[a] The aerosol dose does not include the contributions by the humidifier.

air concentrations tend to be more constant and change more slowly. For this compound, the amount of time spent at home is the most important behavioral factor. This factor tends to be normally distributed in the population, yielding exposure distributions for inhaled vapor of methyl parathion tending toward normal, as shown in Figure 7.14.

The inhalation distribution of the aerosol dose also tends toward lognormal, being skewed at the low end by individuals who do not shower, and thus have minimal aerosol generation in their homes. Chromium, which is nonvolatile, had aerosols resulting from showering as the only source for the inhalation exposure, which was much greater for the individual who showered than the individual who did not, since aerosol penetration through the home originating from showers is small.

The distributions of dermal dose are bimodal for each of these compounds, as the portion of the population that bathed or showered received much greater dosages than the portion of the population that did not bathe or shower on the sampled day. This can be seen in the distributions shown in Figures 7.15, 7.16, and 7.17. For each of the compounds, the dermal dose is increased by approximately a multiple of 30 by the showering or bathing activity.

For each of the compounds studied, showering and bathing were the largest contributors to the dermal dose in the portion of the population who engaged in these activities, due to the longer contact time. Since behavior relating to dermal exposure, and thus skin contact time, was identical for each compound and individual, the relative differences in dermal dose are a result of differences in the skin permeation coefficients. In the case of chloroform, which has the highest skin permeation coefficient of the compounds examined, the dermal exposures were the highest. For methyl parathion, a semivolatile compound with a smaller permeability coefficient than chloroform, the overall dermal exposures were lower than for chloroform. Chromium has a low skin permeability and therefore resultant dermal exposures were smaller than for the other compounds.

A commonly used rule of thumb estimates total exposure at two to three times that of ingestion exposure. This may be a reasonable estimator for a few cases, but it does not adequately account for the variability in the population and the complex interaction of many processes. This estimation technique clearly both underestimates and overestimates exposure, depending on the activities of the population being evaluated. The ingestion exposure for each compound is calculated based on the water concentration, amount ingested, and uptake through the gastrointestinal tract. Assuming complete absorption for each compound and an ingestion rate of 1.4 L/day for 1 mg/L water concentration, the ingestion dose would be 1.4 mg for each compound. The following analyses of the importance of each route are made with the assumption that the ingestion dose is 1.4 mg/day. It is recognized that there is variability in the amount ingested and that all ingested mass may not be absorbed.

If the criteria were to protect 99% of the population from a given level of exposure, the total exposure for chloroform would be the sum of 5.9 mg/day (Table 7.17) and the ingestion dose. Given the above assumption, this would result in a total exposure of over five times the ingestion exposure, indicating that the rule of thumb mentioned above (two to three times) greatly underestimates the actual total dose.

For the single day simulated, the inhalation and dermal doses for chloroform (which is very volatile and has a high permeability coefficient) were both greater than the ingestion dose for the subject who showered and comparable for the other subject, who did not shower. However, for methyl parathion, the inhalation dose was considerably smaller than either the dermal or the ingestion dose. For both subjects, the estimated dermal dose by the second uptake model (see Section 7.6.3.2) was approximately 10% and 30% of the estimated ingestion dose, for the individual who did not shower and the individual who showered, respectively. For chromium, the inhalation and dermal doses were both small compared to the ingestion dose.

The simulations undertaken in this case study can be used to illustrate approaches to evaluating the relative strengths of sources for each of the three compounds. It should be recognized, however, that these discussions should be used only as a guide to how the analysis might be done since this particular study is not comprehensive.

7.8 CONCLUSIONS

The objective of this case study is to demonstrate the methods recommended in this document for assessing exposure and dose as a result of contact with waterborne contaminants. Generally, the models and techniques are well defined, but large uncertainties currently exist in many variable parameters, leading to much uncertainty in the results. In addition, significant natural variability exists as a result of varying physical environments and behavior of individuals in the environment. This case study demonstrates the complexity of factors influencing exposure to waterborne contaminants.

The relative importance of each of the considered uptake routes is summarized in Table 7.17, displaying the estimated uptake and fraction of total by each of the considered routes. The inhalation route is demonstrated to be the primary route for higher-volatility compounds (e.g., chloroform), and is also an important route for the semi- and lower-volatility compounds. The relative importance of the dermal route increases as volatility decreases. The importance of the dermal route is greatly impacted by the chemical's K_{ow} value. Chloroform and methyl parathion have similar values for K_{ow}, resulting in similar estimates of dermal uptake. If a compound with a much higher K_{ow} was investigated, it is likely the dermal route would be of greater relative importance.

Although ingestion uptake was not formally estimated, a comparison with the maximum likely dose provides insight into its relative importance. Given the assumptions discussed in the previous section (resulting in an estimated ingestion dose of 1.4 mg for each compound), the ingestion and dermal routes are of relatively lesser importance for highly volatile compounds compared to inhalation. Furthermore, typically approximately 30% of volatile compounds, such as chloroform, is released into the air prior to consumption (Prichard and Gesell, 1981), thereby further reducing the significance of the ingestion route for volatile compounds.

Alternatively for low- and nonvolatile compounds, ingestion has the potential of being the most consequential route. For example, for methyl parathion, ingesting 1.4 liters of unprocessed tap water would result in a daily absorbed dose more than 30 times that of the combined dose for the dermal or inhaled routes for the 50th percentile individual, assuming complete absorption.

These results emphasize the need to understand the importance of each exposure route associated with waterborne contaminants when estimating the dose. This is especially important when providing advice to individuals or communities on safe uses of their drinking water supply when it has been contaminated. A single set of guidelines will not be adequate, but rather each compound needs to be considered independently along with how the specific community might use the water.

Parameter values used in this case study have been gathered from a variety of sources. Many of these parameters (e.g., characteristics of household appliances, housing characteristics, behavior, etc.) have changed over the years. An exposure assessor should use the most appropriate values for these parameters. For example, the appliance characteristics should be representative of the appliances in the residence or set of residences being studied. Also, specific, regional, national aspects of residential characteristics have a large impact on an exposure assessment (Murray and Burmaster, 1995) and should be carefully considered. Likewise, characteristics specific to the population group (e.g., age-related and other differences in body weight, body surface area) are important and should be considered for a particular assessment.

A significant example of the importance of properly representing input parameters was noted in this case study by the poor representation of bathing and showering activities. As described earlier, the shower frequency specified as 0.7 showers per day per person was not achieved in the simulations due to inadequacies in the activity pattern database used to represent the population's activities, resulting in a simulated frequency of approximately 0.4 showers per day. This has a pronounced effect on the resulting inhaled vapor, inhaled aerosol, and dermal exposures. It is likely that this case study significantly underestimates these exposures for approximately 30% of the population.

It is generally agreed that mass-balance indoor air quality models yield reasonable estimates provided that values of the input parameters are well defined. As many of the model parameters have considerable uncertainty, a formal sensitivity analysis would be useful in identifying the parameters where the acquisition of additional information would lead to the greatest marginal improvement. Furthermore, a well-designed validation study, with the objective of collecting necessary information for validating each of the submodels, as well as the overall exposure estimates, would greatly add to the confidence in these types of models.

Properly validated and applied indoor air quality/exposure models provide estimates of population exposure characteristics and a tool for evaluating remedial strategies at a fraction of the cost of field studies aimed at gathering the same estimates. In addition, a more complete understanding of factors influencing exposure may result in more targeted remedial strategies and more effective use of resources.

The modeling techniques demonstrated in this case study allow a more complete understanding of the interaction of the various factors leading to exposure, and provide useful insight into the important causes, the affected populations, and potential remedial actions. This report and case study emphasizes the importance of estimating the distribution of expected doses for a population rather than several points whose place in the distribution is unknown.

REFERENCES

Axley J., Progress toward a general analytical method for predicting indoor air pollution in buildings: *Indoor Air Quality Modeling Phase III Report*, NBSIR 88-3814, National Bureau of Standards, National Engineering Laboratory, Center for Building Technology, Building Environment Division, July 1988.

Baranowska-Dutkiewicz, B., Absorption of hexavalent chromium by skin in man, *Arch. Toxicol.* 47, 47, 1981.

Berne, R.M. and Levy, M.N. (Eds). *Physiology*, Third Edition. Mosby-Yearbook, St. Louis, 1993, 494–509.

California Air Resources Board, Activity patterns of California residents, *Research Division, California Air Resources Board, Final Report*, Contract No. A6-177-33, May, 1991.

Cleek, R.L. and Bunge, A.L. A new method for estimating dermal absorption from chemical exposure, General approach, *Pharm. Res.* 10, 497–506, 1993.

Gordon, S., Callahan, P., Brinkman, M., Kenny, K., and Wallace, L. Effect of water temperature on dermal exposure to chloroform, 7th Annual Meeting of the International Society of Exposure Analysis, 1997.

Gunderson, E.C. and Witham, C.L., Determination of inhalable nonvolatile salts in a shower, SRI International, Menlo Park, California, SRI Project No. PYC-6631, 1988.

Hakkinen, P.J., Personal Communications, 1997.

Highsmith, N.R., Hardy R.J., Costa, D.L., and Germani, M.S., Physical and chemical characterization of indoor aerosols resulting from the use of tap water in portable home humidifiers, *Environ. Sci. Technol.,* 26, 673–679, 1992.

Highsmith, N.R., Rodes, C.E., and Hardy R.J., Indoor particle concentrations associated with use of tap water in portable humidifiers, *Environ. Sci. Technol.,* 22, 1109–1112, 1988.

Hinds, W.C. *Aerosol Technology: Properties, Behavior, and Measurement of Airborne Particles*, New York: John Wiley & Sons, 1982.

Johanson, Modeling of respiratory exchange of polar solvents, *Ann. Occup. Hyg.* 35, 323–339, 1991.

Keating, G.A. and McKone, T.E., Measurements and evaluation of the water-to-air transfer and air concentration for trichloroethylene in a shower chamber, *Modeling of Indoor Air Quality and Exposure, ASTM STP 1205,* ed. N.L. Nagda, Philadelphia, PA: American Society for Testing Materials, 1993, 14–24.

Mackay, D. and Shiu, W.Y. A critical review of Henry's Law constants for chemicals of environmental interest, *J. Phys. Chem. Ref. Data*, 10(4), 1175–1199, 1981.

Morgan, M.G. and Henrion, M., *Uncertainty*, New York: Cambridge University Press, 1990.

Murray, J.F., *The Normal Lung*, 2nd edition, W.B. Saunders, Philadelphia, 1986, 152.

Murray, D.M. and Burmaster, D.E., Residential air exchange rates in the United States: empirical and estimated parametric distributions by season and climatic region, *Risk Anal*, 15, 459–465, 1995.

Nakai, J., Chu, I., and Moody, R.P., Dermal absorption of chemicals into freshly-prepared and frozen human skin. Originally presented at Gordon Research Conference on Barrier Function of Mammalian Skin, Proctor Academy, NH, and provided to us by personal communication, 1995.

Prichard H.M. and Gesell, T.F., An estimate of population exposures due to radon in public water supplies in the area of Houston, Texas, *Health Physics,* 41, 599–606, 1981.

Selleck, R. E., Mariñas, B. J., and Diyamandoglu, V., Treatment of water contaminants with aeration in counterflow packed towers: theory, practice and economics. University of California, Berkeley, CA. UCB/SEEHRL Report 88-3/1, 1988.

U.S. Department of Housing and Urban Development (U.S. DHUD), Residential Water Conservation Projects: Summary Report. Washington, D.C. Report Number HUD-PDR-903, 1984.

U.S. Environmental Protection Agency (EPA), *Exposure Factors Handbook*, EPA/600/P-95/002Fc, Washington, D.C., 1997.

Vecchia, B.E., Estimating dermal absorption: Data analysis, parameter estimation and sensitivity to parameter uncertainties, Master's thesis, Colorado School of Mines, Golden, Colorado, 1997.

Wilkes, C.R., *Modeling human inhalation exposure to VOCs due to volatilization from a contaminated water supply,* Ph.D. Dissertation. Department of Civil Engineering, Carnegie-Mellon University, Pittsburgh, PA, 1994.

Wilkes, C.R., Small, M.J., Davidson, C.I., and Andelman, J.B., Modeling the effects of water usage and co-behavior on inhalation exposures to contaminants volatilized from household water, *J. Expos. Anal. Environ. Epidemiol.*, 6(4),393–412, 1996.

Yeh, H.C. and Schum, G.M., Models of human lung airways and their application to inhaled particle deposition, *Bull. Math. Biol.* 42, 461–480, 1980.

Index